The Zero-Point Universe

By

Ray Fleming

This publication was published, written, edited and illustrated by Ray Fleming, with additional editing by Ward Lowe.

Published in the United States of America

ISBN-10: 1470033976
ISBN-13: 978-1470033972

CONTENTS

Acknowledgement

Thanks to Dinu Popa, Robert Tisdale and Wayne Friesell for the many conversations over the past few years while I developed the theories in this book.

Introduction: The Genesis of a Physics Heretic

That one body may act upon another at a distance through a vacuum, without the mediation of any thing else, by and through which their action and force may be conveyed from one to another, is to me so great an absurdity, that I believe no man, who has in philosophical matters a competent faculty for thinking, can ever fall into it.[1]

Sir Isaac Newton, 1693

Did you ever get the feeling that physics was just too contorted to be correct? I did. I enjoy studying the way things work and being able to predict results with simple mathematical equations. However, I found the whole idea of action at a distance unacceptable, and it was unfortunately standard to all accepted force theories. When I began to study more complex physics like General Relativity, Quantum Mechanics, Quark Theory, and String Theory, my first thoughts were, *What the heck is this...stuff?*

Curved space, uh, that makes no sense. Ninety percent of the matter is missing. Now if I was taking a test and I got a problem 90% wrong, I would fail. Even if I were only 70% wrong, I would still fail. Quantum Mechanics is nice, but what about the force that opposes electrostatic attraction that prevents an electron from falling into a proton. Hey guys, come back here, you forgot something. Then there is Quark Theory; you know, I read a book once on how measurements in the Great Pyramid correspond to dates of important events in time. It was very interesting stuff. Next up String Theory; but I thought physics theories were supposed to describe something, well, physical - silly me.

Perhaps the best one is that force interactions are supposedly relayed between bodies of matter by gauge bosons such as photons for the electro-magnetic forces. Now these gauge bosons are really smart particles. They have something like a memory chip in them that can store information, such as what type of charge they came from, so other bodies can tell whether they need to be attracted or repelled. It also has to record how much charge, so the other body knows how much to be attracted or repelled. Then it stores information on how fast it was going and what magnetic field was produced in the process. It does that by knowing where the universal rest frame is so comparative velocities can be computed when it hits another body. Oh, I forgot there is not supposed to be a universal rest frame to allow translations of measurements. But assuming there is one, when a smart gauge boson hits a body, the body has a small supercomputer with billions of inputs, one for each and every gauge bosons that hits it every zeptosecond. The supercomputer makes the amazingly complex calculations to determine what it has to do next. Then the body deploys a set of mini jet engines, pointing them in the proper direction, burning the right amount of fuel, so the body moves the way it is supposed to.

Of course that last paragraph is complete nonsense, as is any gauge boson theory. Gauge bosons are smart particles that supposedly transmit forces, but in reality they have no memory chip, and the bodies of matter have no supercomputer. Gauge bosons do not have any fundamental properties that would allow them to store such detailed information and certainly no way to identify a comparative frame of reference to allow them to determine relative velocities. The smart particle theory also fails to include a mechanism for how bodies are supposed to move after getting this information. The smart particle transmitting force idea (i.e. gauge boson theory) is incredible, as in "not credible." Whatever it is that transmits forces through space must work in a simpler, more fundamental way.

Most of my younger life I had wanted to "grow up" to be a physics professor. I wanted to get my PhD and teach at a fine university and do my best to figure out all I could about the universe. I was dumbstruck over how senseless the Standard Model of physics was. So instead, I decided to fall back on plan B, to be a photographer. Well I will tell you, breaking into professional photography is really tough, and so, I went back to physics. But, when the time came to pick a graduate school, I decided I could not risk it. I felt that by exposing myself further to the theories of 1980s, I might permanently destroy my innate creative ability to come up with better theories in the future. And so, my long trek as an independent researcher began.

In my mind, the ultimate theory of the universe had to be based fundamentally on a mechanical model. Each and every force interaction must be due to a force mechanism, which allows point-to-point transmission through space. And, once transmitted, there needs to be a basic property of space that causes bodies to move, with no jet packs required. This philosophy of physics is not something new. It was the standard in the mid-1500s up to the time when Newton published his *Principia*. The strongest advocates of a mechanical theory of that time were two of the fathers of modern science, Galileo Galilei and René Descartes. In the Cartesian view, space is filled with vortices that respond to and cause movement of the planets. These vortices were responsible for gravity and other orbital interactions. Prior to Descartes, a view of an æther filled with corpuscles was more common, but the two theories coexisted.

When Newton published his *Principia* all that changed. He had made substantial improvements to the mathematical model that describes the movement and interaction of bodies. At the same time he failed to present a mechanical cause behind those interactions. He simply chose not to try. At one level this was smart, as he simply did not have enough information to get the theory correct. But more importantly, it set a bad precedent. It was now OK to propose models and theories that did not account for an action mechanism. Magical force transmission theories became the norm. The ideal of a working mechanical force theory was lost.

Actually, it was not lost entirely. Nicolas Fatio de Dullier a contemporary of Newton's, proposed the push theory of gravity.[2] He based his theory on hypothetical corpuscles of the vacuum pushing bodies together. They screened each other from the pressure that would otherwise push them apart, making the force pushing them together greater. This theory is brilliantly simple, both in its methods of force transmission and interaction. Analysis showed that the theory was not correct in the form proposed by Fatio, but we will look at this theory in later chapters. A number of brilliant physicists, including Lord Kelvin, James Clerk Maxwell, Hendrik Lorentz, J.J Thompson, Henri Poincaré, and Richard Feynman, chose to examine Fatio's Theory, also known as Le Sage's Theory since Le Sage was responsible for popularizing it. The desire to find a physical mechanism behind gravity was not entirely lost; it only felt like a lost cause because the nature of the corpuscles was misunderstood.

Everything changed for me in the fall of 1990 when a coworker reintroduced me to zero-point energy theory in the work of Hal Puthoff.[3] The idea of vacuum fluctuations had come up during my undergraduate work, but not in terms of being something that we could seriously think was responsible for force interactions. Puthoff's paper on the hydrogen atom was not convincing to me since it failed to address the force opposing electrostatic attraction. But it was plain to see that in zero-point energy I had found what was missing from the other physics theories. This was something real that actually filled the vacuum. It is the only thing we know that fills the vacuum of space. I then knew that zero-point energy had to be responsible for the mechanisms behind every force. I was not alone in thinking this way. As the Internet grew, it was soon filled with many other searchers like me, all thinking along the same basic lines; and there was much history as well.

On the other hand, I believe every physicist feels inclined to the view that all the forces exerted by one particle on another, all molecular actions and gravity itself, are transmitted in some way by the ether, so that the tension of a stretched rope and the elasticity of an iron bar must find their explanation on what goes on in the ether between the molecules. Therefore, since we can hardly admit that one and the same medium is capable of transmitting two or more actions by wholly different mechanisms, all forces may be regarded as connected more or less intimately with those which we study in electromagnetism.[4]

Hendrick Lorentz, 1906

Mainstream physicists, however, still treated zero-point field theory like it was a curiosity, which was not useful, or even worse, a joke. Unfortunately, they had some good reasons to see it as a joke. Some zero-point energy searchers had glommed onto the idea that zero-point energy was a source of free energy. Some brought it into their theories of mysticism or metaphysics. Many amateur papers on the Internet could never survive peer review, and free energy experiments were not reproducible. It was also tied in scientists' minds to discredited theories of the vacuum from the past, the static æther theories, which were thought to behave in accordance with kinetic gas laws. It is little wonder physicists dismissed zero-point field theory with a laugh, much as I had dismissed some of their theories a few years before.

Over time I worked on the puzzle, and a difficult and challenging puzzle it was. The key was to avoid being sloppy; to steadfastly adhere to basic scientific principles with the most important being that forces must be transmitted by the zero-point field. I am now at the point where I can see where pieces fit without looking at them, as the answers appear obvious and everything simply falls in place. The beauty, the elegance, and the simplicity of the theory is amazing, and I want to share it. For in the zero-point universe, everything can be described by a single simple mechanical theory.

[1] I. Newton, excerpt from a letter to Dr Bentley dated 25 February 1693.

[2] N Fatio de Dullier, "De le Cause de la Pesanteur" (ca 1690), Edited version published by K. Bopp, Drei Untersuchungen zur Geschichte der Mathematik, Walter de Gruyter & Co. pg 19-26 1929.

[3] H. E. Puthoff, "Ground State of Hydrogen as a Zero-Point-Fluctuation-Determined State," Phys. Rev. D 35, 3266 (1987).

[4] H. A. Lorentz, The Theory of Electrons, 2nd Edition , B. G. Teubner, Leipzig; G. E. Stechert & Co., New York, 1916, p45. (Taken from a series of lectures presented in 1906.)

Chapter 1: The Philosophy of Science

My own opinion is that we ought to search for a way of making fundamental changes not only in our present Quantum Mechanics, but actually in Classical Mechanics as well. Since Classical Mechanics and Quantum Mechanics are closely connected, I believe we may still learn from a further study of Classical Mechanics. In this point of view I differ from some theoretical physicists, in particular Bohr and Pauli. [5]

Paul Dirac, 1951

The Closing of Theories

In the quote above, Paul Dirac is addressing the worst thing that ever happened to science: the concept of closed theories. Throughout his life Dirac was the champion of the idea that all theories should be open for further evaluation or interpretation. Werner Heisenberg, on the other hand, was the leader of a movement starting in the 1930s and continuing throughout his life that treats certain theories as being forever closed.[6] Other important physicists such as Bohr and Pauli joined with Heisenberg in his ideals. To Heisenberg, a closed theory is one that is "perfectly accurate within its domain" and "correct for all time."[6] He went on to champion four theories for closure: Newtonian Mechanics, Thermodynamics, Electricity and Magnetism (including special relativity and optics), and Quantum Mechanics. He conspicuously omitted General Relativity from his personal list, but there were plenty of other people to champion it as a closed theory without him.

The closed theorists could not have had better timing, as the movement was gaining strength just as World War II was looming. Many physicists had to break away from their pursuits of fundamental physics theory in order to pursue survival, to find a new country free of persecution, or to locate a new position with political ties that would allow them to safely weather the coming war. Then during the war, many great physicists were drafted into weapons development programs and became managers, engineers, and technicians. By the time the war ended, more than a decade had passed with no substantive changes in fundamental physics theory. After that, the obedient young solders returned to the classroom to learn physics just as it was taught, and for the most part fully accepting the notion that the earlier theories were closed. This continued for decades. The closed theorists had won. Science, and indeed the world, had lost.

> I feel very strongly that the stage physics has reached at the present day is not the final stage. It is just one stage in the evolution of our picture of nature, and we should expect this process of evolution to continue in the future, as biological evolution continues into the future. The present stage of physical theory is merely a steppingstone toward the better stages we shall have in the future. One can be quite sure there will be better stages simply because of the difficulties that occur in the physics of today.[7]
>
> Paul Dirac, 1963

Well after the war, Dirac was still unwilling to give up, as he recognized serious deficiencies in the state of physics. He went on to say that even his theory of the electron was suspect, and worried that it might ultimately be found to be incorrect. Unlike the others, Dirac was true to the scientific method. All theories are continually open to revision, even his own. And given sufficient compelling evidence, an old theory must be discarded and replaced by a new theory.

Class-One or Class-Two Problems
Dirac broke the problems of physics into two classes. He viewed class-one problems as those that have to deal with the "consistency with nature." He goes on to take the unfortunate position that "it is only the philosopher, wanting to have a satisfying description of nature, who is bothered by class-one difficulties." "If the physicist is able to calculate results and compare them with

experiment, he is quite happy if the results agree with his experiments, and that is all he needs."[7] I disagree, and I guess that makes me a physicist-philosopher. To me, someone who is only interested in making computations that match observations is little more than an engineer.

A. Class-one problems are those that require a new fundamental theory.

B. Class-two-problems are those where the existing fundamental theory seems satisfactory, and the problem within that theory appears to be solvable without modifying the existing theory.

Certain problems cannot be solved within the scope of the existing theory and are truly class-one problems. Important examples of class-one problems are the cause of inertia, Dark Energy, the missing matter problem, the force opposing gravity that keeps a spinning top from falling, and the opposing force between an electron and proton. It is up to physicists and other scientists to identify the class-one problem theories, accept that problems truly are class-one, and come up with better replacements.

With so many of the basic theories being considered closed, the vast majority of work in physics over the past 75 years has been at the class-two level. That is a sign that perhaps there are too many physicist-engineers and not enough physicist-philosophers. That is not to imply that the work of the physicist-engineer is any less important, as the solution of class-two problems is important to the advancement of science and equally important to the transition of ideas from pure science concepts to the improving designs of things mankind can use.

The catch is how do you know if a problem is class-two or class-one? Closed theorists will obviously throw all problems into the class-two category, with the mentality that only the existing theory provides the proper framework. Then they will deny that the theory is a failure if their attempts to fix it do not succeed. A physicists with the mindset of Dirac may simply decide not to be bothered if he feels that the objections are of a philosophical nature and therefore not important to him. These two factors alone make it highly unlikely that scientists will re-categorize class-two problems as class-one. Not even time seems to make a difference. The missing matter problem has been around ninety-some years and gravitational theory is somehow still not seen as a serious problem. At least it is not a serious enough problem to raise it to class-one, and consequently declare that General Relativity needs to be replaced. Time needs to be taken into account, although it certainly is not the only measure. Class-two problems sometimes take decades to solve. But, if an entire generation of physicists lives out their lives from beginning to end with an unresolved class-two problem, that is a good indication that the entire theory must be re-evaluated as a class-one problem.

Ultimately, the only sure way to raise a problem, and thus a theory, to the status of class-one, is for someone to come out with a better theory. Someone or a group of people has to see the problems and work on new theories, possibly outside the mainstream, in the hopes of coming up with something better. This unfortunately leads to an us versus them situation, the heretics versus the true believers. Once again, science suffers as sides are taken without regard to evidence or unbiased evaluations of the competing theories.

Is it Science or Science Fiction?
Anyone who watches cable television programming or reads popular science magazines, blogs, or mainstream media articles will find there is a blurring of the lines between science and science fiction. Science fiction is exciting and it sells, while real science tends to be much more boring. People and even some scientists would rather speculate on parallel universes, faster-than-light travel, and wormholes rather than stay within the confines of physics based on real observations and scientific experiments.

After closed theories, the influx of science fiction into science is the second biggest problem with today's physics. The problem is even greater than most people realize, as it has inundated many parts of physics, such as curved space-time, length contraction, extra dimensions, quarks, strings, or anything with relativity in its name. Mathematicians are frequently to blame as they somehow feel that an interesting mathematical problem must have a physical analog, and in so doing promote science fiction as science. This is nothing new, as it has been going on for more than a century.

The field of zero-point energy research is unfortunately one of the worse fields of physics for science fiction being reported as science, while at the same time the real science is dismissed as science fiction. In this book, I will attempt to limit the scope to the science of the zero-point universe by ignoring the science fiction elements. We will find that any possible theory that is simple enough and comprehensive enough to be capable of unifying all the forces will ultimately be boring and not resemble science fiction in the least.

> *The only equation by which the observed phenomena are satisfactorily accounted for is that of Planck, and it seems necessary to imagine that, for short waves, the connecting link between matter and ether is formed, not by free electrons, but by a different kind of particles, like Planck's resonators, to which, for some reason, the theorem of equipartition does not apply. Probably these particles must be such that their vibrations and the effects produced by them cannot be appropriately described by means of the ordinary equations of the theory of electrons; some new assumption, like Planck's hypothesis of finite elements of energy will have to be made.*[8]

Hendrick Lorentz. 1906

Fundamental Principles

The consequence of all the science fiction being mixed in with the science is that physicists have become sloppy when it comes to the fundamentals. As we will see as we progress through the book, as long as we stay consistent with the fundamental principles of physics the correct answers and theories will reveal themselves.

As a guide for both heretics and true believers alike, here is a list of fundamental principles to guide us as we evaluate both existing and proposed theories to try to find those with an underlying realistic mechanical explanation. Oh, and do not worry if any of these concepts are new to you. They will be explained and examined in greater detail in coming chapters.

1. Every theory is open

The scientific method must not allow closed theories. Every theory must be open for discussion and must be able to be brought into question, always.

2. Treat all problems as class-one problems first

The present theories have drifted so far from the Cartesian ideal of having a mechanism behind each force theory that every theory must be called into question. Considering problems as class-one first forces us to examine a greater range of possibilities, and hopefully dismiss old theoretical prejudices in the process. Keep in mind, though; the problems may end up being class-two problems after all, with only the mechanism requiring explanation. In rare instances, the theory may be essentially correct as is.

3. Science fiction is not science, or not every mathematical equation has a physical interpretation

One of the saddest things about being a physicist today is seeing how much science fiction is being taught as real science. This problem came about largely because certain mathematicians adopted the view that every mathematical concept has a real physical analog. That view is unnecessary and not scientifically supportable. One of the worst examples is the treatment of mathematical dimensions as physical dimensions. Just because you solve an N variable mathematical problem in an N dimensional mathematical problem-solving space, does not mean that those N dimensions are physically real. Science fiction may be fun in movies, but there needs to be a firm boundary between it and real science.

4. Every force is transmitted by and through the zero-point field

We must do away with or modify all theories that rely on magical force transmission. Based on the present knowledge of physics there is only one thing that permeates the vacuum: zero-point energy. This collection of zero-point energy, the zero-point field, is the only thing that can provide the mechanisms to explain force transmission through the vacuum. These mechanisms must be simple, not requiring numerous bits of memory and a supercomputer to process the data as a gauge boson theory would. If a theory does not already have an underlying mechanism consistent with zero-point field theory, we will need to replace it with a new theory.

5. Every theory must adhere to Heisenberg's Uncertainty Principle

As we will see, it is too frequently the case that physicists forget that their theory must be in accordance with the Uncertainty Principle or more often the inverse of the Uncertainty Principle. In real terms, this means that they theorize about particles existing longer than they are allowed to without violating the inverse Uncertainty Principle, since zero-point events cannot be directly measurable. Such theories must be modified or disposed of.

6. Every theory must adhere to the principles of conservation of energy and momentum

Violations of the uncertainty principle often lead to violations of conservation of energy and momentum. These theories and any others that violate conservation of energy and momentum likewise need to be modified or done away with. There is, however, one major difference between Standard Models theories and the theory of the zero-point universe, and that is that potentials in old theories are zero-point field pressure in the new theories. That pressure can be thought of as energy due to the zero-point field.

> *All of our experience, without a single exception, enforces the proposition that no body moves in any direction, or in any way, except when some other body in contact with it impresses its own motion upon it. [...] For mathematical purposes, it has sometimes been convenient to treat a problem as if one body could act upon another without any physical medium between them; but such a conception has no degree of rationality, and I know of no one who believes in that as a fact. If this be granted, then our philosophy agrees with our experience, and every body moves because it is pushed, and the mechanical antecedent of every kind of phenomenon is to be looked for in some adjacent body possessing energy; that is, the ability to push or produce pressure.[9]*
>
> Amos Emerson Dolbear, 1897

7. Everything moves because it is pushed
And even at a state of rest everything is being pushed. The concept of an attractive force is an illusion. Zero-point vacuum fluctuations do not come with hooks. They cannot pull on a body. If two bodies move toward each other due to a force it is because the pressure pushing them together is greater than the pressure pushing them apart.

8. Space is free of prior structure
Space does not have some kind of primordial lattice structure or any other type of prior structure. Any theory of space must start out with something geometrically flat. The underlying space is not contracted, time dilated, or curved. Any model that starts with prior structure is incorrect.

Science alone of all the subjects contains within itself the lesson of the danger of belief in the infallibility of the greatest teachers in the preceding generation . . . Learn from science that you must doubt the experts. As a matter of fact, I can also define science another way: Science is the belief in the ignorance of experts.[10]

Richard Feynman

9. Set aside hero worship or negative feelings about individual scientists

This is not about individuals; this is about science. Historically, some scientists' theories have been more readily accepted due to their stature in the physics community or with the media. Other scientists have been dismissed for irrational reasons or reasons unrelated to the quality of their theories. Those feelings about individuals have to be set aside in order to make a full and independent evaluation of all theories.

10. Start with a logical and intuitive model

A mathematical model is important, but it is easy to hide overly complicated or even erroneous ideas in mathematical equations. To put it simply, if a theory introduces a paradox or is self-contradictory then it is incorrect. Theories must be logically consistent, so if contradictions cannot be resolved, the theory must be discarded. If an idea is simple enough to be fundamental, it is simple enough to be described in a simple, logical, intuitive, non-paradoxical manner.

11. Beware of numerology

It is not all that uncommon to throw numbers and equations together in random ways in the hopes that the correct result will magically appear. Scientists have been doing this with the Fine Structure Constant $(1/{\sim}137)$ for nearly a century. A numerologically derived answer in the form of an overly complex equation thrown together without an underlying logical reason is seldom correct.

12. Theory must match observation

Being a basic tenant of science, you would think this would be easy to comply with, but it is not. It is sometimes the case that experiment contradicts theory, but physicists have become so enamored with the theory that they refuse to dismiss it. This means more than just having equations that produce the correct answers. All parts of the theory must be taken from physical reality, not a fantasy world. If something is not known to physically exist, without even an inkling that it might exist - it has no place in a theory. On the other hand, if an experimental result means that a theory is untenable, then the theory must be discarded. This should not be difficult, but for some physicists it is. They would rather live in a fantasy universe than the real one.

Conclusion

While we could add more to this simple list, those rules will be sufficient to prevent us from being as sloppy as our predecessors. This list of fundamental principles will guide us as we search for a purely mechanical understanding of the universe. In the process it will help us discard the magical force theories of the past. By adhering to these principles we will gradually come to know the true nature of the zero-point universe.

[5] P.A.M. Dirac, The Relation of Classical to Quantum Mechanics. Proceedings of the Second Mathematical Congress, Vancouver, 1949. Toronto: University of Toronto Press. Pg. 18 (1951).

[6] A. Bokulich Open or Closed? "*Dirac, Heisenberg, and the Relation between Classical and Quantum Mechanics,*" Studies in History and Philosophy of Modern Physics 35(3) (2004)

[7] P.A.M. Dirac, *The Evolution of a Physicist's Picture of Nature*, Scientific American 208:5 pp 45-53 (May 1963)

[8] H. A. Lorentz, The Theory of Electrons, 2nd Edition , B. G. Teubner, Leipzig; G. E. Stechert & Co., New York, 1916.

[9] A.E. Dolbear, Modes of motion; or, Mechanical conceptions of physical phenomena Boston, Lee and Shepard (1897)

[10] R. P. Feynman, The Pleasure of Finding Things Out p. 186-187, 1999.

Chapter 2: The Zero-Point

> *The enormous factor from nuclear densities of ~10^{14} g/cm³ to the density of field fluctuation energy in the vacuum, ~10^{94} g/cm³, argues that elementary particles represent a percentage-wise almost completely negligible change in the locally violent conditions that characterize the vacuum. In other words elementary particles do not form a really basic starting point for the description of nature.* [11]
>
> John Wheeler & Charles Misner, 1962

What is the Zero-Point?

To understand the physics of our universe we must begin at the zero-point. What is the zero-point? To begin, first imagine a box filled with nothing but air. Then remove all the air so there is not a single atom or particle in it. After that, shield it from all waves of light energy and electric or magnetic fields and reduce the temperature to absolute zero. When we are done, the space inside the box will be at the zero-point, a state of absolute nothingness. At that point we cannot see or detect anything at all inside the box. Early scientists treated the zero-point as if there was truly nothing there, as if it was absolute emptiness, a null space. It did not play a roll in our understanding of the universe, or in their equations, other than as a backdrop. Unfortunately many scientists today, even though they know better, still treat the zero-point that way.

Nearly a century ago, physicists recognized that the zero-point was more than a vacuum with nothing in it, but rather it is teeming with energy. What they discovered is that there can never be an absolute zero, an absolute nothingness. There are always tiny vibrations, vacuum fluctuations, present. The amount of energy in the fluctuations over the time they exist is so small as to be undetectable. But while each fluctuation is small, the vacuum is teeming with them, and when one goes away another takes its place. Taken altogether the amount of energy at the zero-point is unfathomably huge. Instead of our box at the zero-point having no energy, it actually has much more energy than any normal space filled with matter.

The most commonly quoted mass for the vacuum is 10^{94} **grams per centimeter cubed (g/cm^3)** as calculated by John Wheeler who was quoted above.[11] We will calculate later that the energy of the vacuum is 10^{95} **g/cm^3** by a slightly different method, so that value will be used from here on. As we will see, one order of magnitude difference is not that significant at this point in our discussions. Energy is related to mass by the well-known relation **E=mc^2**. For comparison water has a mass density of **1 g/cm^3** by definition. It is impossible for most normal people to grasp just how big a difference in energy there is between the zero-point field and water, so perhaps a simple illustrative example will help.

Let's start with the clichéd drop in a bucket. If the drop is one milliliter (**1 ml**) and the bucket **100 liters** (**72.5 gallons**), then that gives us a factor of 10^5. If instead we consider a drop in all the Earth's oceans, then we have a factor of 10^{24}. That is a lot bigger than a bucket but nowhere close to how insignificant the mass of the drop of water is when compared to zero-point energy. To continue, what if the ocean was the size of the sun? That gives us a ratio on the order of 10^{41}, which is still a long way off. If the ocean was the size of the solar system we get a ratio on the order of 10^{50}. Now if we expand the ocean to the size of the galaxy we get ~10^{76} and we are still not anywhere close.

What if the ocean is the size of the known visible universe? Assuming a radius of **7.4 x 10^{26}** meters the mass ratio is **5 x 10^{95}**. There we go. So, the density of water compared to the energy of the vacuum is equivalent to five **1 ml** drops of water in an ocean the size of the visible universe. Since we are mostly water and have a similar density to water, the vacuum fluctuations inside our body are like having all the mass-energy of an ocean of water the size of the universe inside each little part of us. Wow, we are pretty insignificant in the big scheme of things and so is any other body of solid matter or any amount of energy associated with it. This zero-point energy is all around us and all throughout us. We are lucky that zero-point energy is not detectable or anything we did would be undetectable noise to any sensor we could possibly make. Even worse, if we could absorb even a small fraction of that energy, we would be vaporized in an instant. Or, if all that energy participated in a gravitational force, the universe would be crushed to a speck.

Max Planck's Oscillators

The history of zero-point energy begins with Max Planck's theory of harmonic oscillators in 1900.[12,13] This was the first appearance of Planck's constant **h**, and is also considered to be the beginning of quantum mechanics. He was the first to describe that oscillating systems related to a black body radiator do not reach a true zero-energy state but rather have a small amount of energy greater than zero as their base state. In his next breakthrough paper "A new radiation hypothesis," he published a more generalized equation as a description of the ground state of the harmonic oscillator as it pertains to black body radiation.[14] He described the energy of the harmonic oscillator with the equation shown as Equation 2-1 where **E** is energy, **ħ** is the reduced Planck's Constant (Planck's Constant **h** divided **2π**) and **ω** is the circular frequency (linear frequency **v** times **2π**).

Equation 2-1

$$E = \frac{\hbar \omega}{2}$$

19

Everything that relates to the zero-point energy of the vacuum relates back to this equation. This equation is the foundation of quantum electrodynamics and the quantum theories of gases. It cannot be overstated how important it is to the field of physics. Whatever importance you put into the equation $E=mc^2$, this is more important. It is the underlying foundation for everything in the universe.

Albert Einstein's Theories of Gases

At the same time, as Planck was working on the black body radiation problem, Albert Einstein was working on a low energy theory of gases making use of Planck's ideas. The first of these papers published in 1910 was coauthored by Ludwig Hopf.[15] The Einstein-Hopf equations are still fundamental to low temperature thermodynamics to this day. The second of those papers by Einstein, coauthored by Otto Stern, was published in 1913 and is remarkable for the first appearance of the German term *Nullpunktenergie,* literally zero-point energy.[16] It is remarkable that Einstein actually shares credit for coining the term zero-point energy. As we shall see it is also somewhat ironic, as Einstein spent so much effort denying the existence of zero-point energy of the vacuum, which lead to many of his theories ultimately being incorrect.

Walther Nernst's Vacuum Fluctuations

> *Even without the existence of radiating matter, i.e., matter heated above absolute zero or somehow stimulated, empty space — or, as we prefer to say, the luminiferous æther—is filled with radiation.*[17]
> Walther Nernst, 1916

In his landmark paper "An Attempt to Return to the Assumption of Continuous Energy Variations from Quantum Theoretical Considerations," Walther Nernst, better known for the Third Law of Thermodynamics, published perhaps the second most important idea in physics after Planck's quantum harmonic oscillator.

Nernst recognized that the vacuum is filled with zero-point energy. Note that Nernst may not have been the first to think of it, and many writers have given that credit to Planck, even though Planck did not publish it first. Perhaps it is Lorentz who should be given credit for his 1906 lecture quoted in the last chapter.[18] But unlike Lorentz, Nernst recognized and explicitly stated that zero-point field theory is equivalent to a luminiferous æther theory. Oh, there it is - the æ word. It is a shame that modern instruction in physics is so poor that people do not realize that æther theory in the form of vacuum fluctuations is necessary to Quantum Electrodynamics and in particular Quantum Field Theory. One form of æther theory was also perfectly acceptable within the scope of General Relativity according to Albert Einstein, although he may not have held that view for long.[19]

After Nernst' made his observation, the fundamental theory of zero-point energy of the vacuum was in place. We had at last found the true nature of the corpuscles that filled space and of Descartes's vortices, although most physicists did not realize it. Nernst went on to speculate on the importance of zero-point energy in terms of a free electron being in equilibrium with the zero-point field, consequently preventing it from radiating away. He also speculated that it was the pressure of the zero-point field that prevented gravity from compressing the universe. Nernst also thought that the zero-point field could absorb the heat and light emitted into space and prevent a heat death of the universe. That sounds like a source of cosmic background radiation. In zero-point energy he also found a mechanism that could account for Einstein's cosmological constant. Perhaps this is why Einstein chose to publicly embrace æther theory not too long after. While his ideas were highly speculative, Nernst recognized that a lot of answers to critical problems could possibly be found by better understanding zero-point energy. Even though his speculations were largely incorrect, this was very advanced thinking for its time.

There is an entire chapter to come later on the discrediting of æther theory and the Michelson-Morley experiment, so we will set the æ word aside for the time being except for the occasional appropriate quotation.

Bose-Einstein Statistics

While not as frequently associated with zero-point field theory, Satyendra Nath Bose's paper "Planck's Law and the Hypothesis of Light Quanta" is the next important addition to the model of the zero-point universe.[20] He recognized that in a case where most photons were zero-point fluctuations in accordance with Plank's quantum oscillators, that they would not behave like a normal gas, but would have different behavior. A collection of photons with equal energy in the same space will behave in the same way and no longer behave as distinct particles. Einstein helped to promote and develop Bose's theory, so it usually carries both their names. While this theory is not mentioned very often in this book, it may be useful at times to think of the vacuum as a Bose-Einstein condensate when considering the transmission of forces through the zero-point field.

The Heisenberg Uncertainty Principle

The next development came about not from the perspective of what happens at zero energy, but rather at what point are we no longer able to measure something. Of course this means the Heisenberg Uncertainty Principle put forth by Werner Heisenberg in 1927.[21] The standard ways of expressing the Uncertainty Principle are either by position and momentum, or time and energy as shown in Equations 2-2 and 2-3.

Equation 2-2

$$\Delta x \Delta p \gtrsim \frac{\hbar}{2}$$

Equation 2-3

$$\Delta E \Delta t \gtrsim \frac{h}{2}$$

When you try to measure the position and momentum of an object, the resultant measurements have a combined error greater than $h/2$ as shown in equation 2-2. It is also stated sometimes that you cannot know the position and momentum of an object at the same time. Equation 2-3 restates the principle with respect to time and energy. There is always some error in knowing the amount of time an event takes and the amount of energy involved. When discussing zero-point energy, the principle is turned around. Zero-point energy is not directly measurable, so the amount of energy of a zero-point fluctuation multiplied by the time it exists must always be less than half of Planck's constant. Comparing Equation 2-1 with Equation 2-3 you can see the similarity. Keeping in mind that $\hbar = h/2\pi$ and the angular frequency ω is the ordinary frequency ($\nu = 1/t$) multiplied by 2π, we get the equivalence shown in Equation 2-4.

Equation 2-4

$$E = \frac{\hbar \omega}{2} = \frac{h\nu}{2} \equiv \Delta E \Delta t = \frac{h}{2}$$

Planck and Heisenberg achieved the same result by approaching the problem from two different directions. What is interesting to keep in mind is that the oscillator energy is at the maximum allowable energy under the uncertainty principle, so with respect to quantum harmonic oscillators, the undetectable ones are always at Planck's minimum energy $h\nu/2$. And they said we could not know the energy. Note that Equation 2-3 is sometimes shown without the factor of $1/2$ such as in Einstein's papers on gases. In those cases both the energy of the oscillator and the energy of the electromagnetic field are both included with each part being *h/2*. We will see that this is a theme throughout physics, that a factor of *1/2* is used when the energy in the zero-point field is neglected.

In physics literature it is not uncommon to see someone refer to zero-point energy being due to Heisenberg's Uncertainty Principle or Planck's principle of the quantum harmonic oscillator. This is not technically correct as it is due to Planck only; however, both approaches are correct in that they work out to the same value. Both interpretations of the fundamental quantum energy will be seen over and over again as we explore the zero-point universe.

> *Furthermore, one might infer that if in any way a portion of the ether were to be annihilated, what was left would at once fill up the vacated space so there would be no record left of what had happened. Apparently its destruction would be a destruction of a substance, which is a very different thing from the destruction of a mode of motion. In the latter only the form of motion need be changed to obliterate every trace of the atom. In the former, there would need to be destruction of both substance and energy; for it is certain, for reasons yet to be attended to, that the ether is saturated with energy.* [22]
>
> Amos Emerson Dolbear, 1897

The Infinity Problem

That completes the basic background of why there is zero-point energy. None of the above science is in dispute, and yet mainstream physicists still resist referring to zero-point energy. The resistance is focused not so much as relates to the theories of gases, including such things as Bose-Einstein condensates that are now experimentally confirmed and widely accepted, but rather on the zero-point energy of the vacuum. The problem with the zero-point energy of the vacuum is not so much in the theoretical principles, but rather that it can be difficult to deal with mathematically. You see, zero-point energy of the vacuum as we have described it so far is infinite, or nearly so, and there is no good may to include something infinite or nearly infinite into an equation. What physicists will do instead is called a renormalization. Basically they make the energy of the vacuum disappear and go on their merry way without it.

Because many current theories in physics require an *ad hoc* deletion of the zero-point energy, many physicists simply prefer not to think about it. Based on the criteria we have set up for our investigation, any theory that requires renormalization to remove the zero-point energy is fundamentally incorrect and will require modification or replacement. That is a lot of theories, many well accepted, so there will certainly be a great deal of resistance to those changes. In the mean time physicists continue to have a two-faced way of dealing with zero-point energy: acceptance on one hand and denial on the other. Their hand will be forced, however, since the Lamb Shift and the Casimir Effect demonstrate conclusively that the zero-point energy of the vacuum is a real physical phenomenon.[23, 24]

Another part of the infinity problem is to determine whether or not the energy of the zero-point field is infinite or only nearly infinite. Or, more importantly, whether a force produced by the zero-point field is infinite or nearly infinite. The problem is that even if a force is diminished somewhat, if it is infinite it is still infinite. If a body pushed on two sides by two infinite forces, there will be no net difference to cause movement. But on the other hand if all infinite forces were actually applied to a body from two sides, it would be crushed to a point. Since we know that things do move due to forces and are not crushed, we know that net forces derived from zero-point energy are less than infinite. We also know that since all matter is not crushed, that the vast majority of the zero-point energy does not act directly on matter kinetically.

One of the first ways to deal with the infinite energy problem was to appeal to the Planck Length. The Planck Length is defined as shown in Equation 2-5 where it is a function of the reduced Planck's Constant $\hbar$, Newton's Gravitation Constant **G**, and the speed of light **c**.

Equation 2-5

$$\ell_P = \sqrt{\frac{\hbar G}{c^3}} \approx 1.616252(81) \times 10^{-35} \text{ meters}$$

The Planck Length comes about when the wavelength of the vacuum fluctuation is at the limit where it can no longer overcome gravity due to the self-energy of the vacuum fluctuation. It is similar to the Schwarzschild radius where the Schwarzschild radius defines the size of a black hole for a body of a given mass. In this case it is energy, not mass, but the same principle applies. A zero-point energy event with a wavelength equal to the Plank Length or smaller is effectively a virtual black hole. Because of that, it cannot participate in interactions limited by the speed of light. It is possible that zero-point energy events this size or smaller are not only undetectable directly; there may not even be a way to detect them due to secondary interactions, as it may be incapable of interacting in any way with matter, other vacuum fluctuations, or anything at all. Whether these higher energy vacuum fluctuations still exist or not is another question, one that we do not, and possibly cannot have an answer to. To even guess at the properties of these virtual black holes we will first need to know the physical structure of the vacuum fluctuations, which is beyond our knowledge at present.

The mass equivalent value for the energy of the vacuum derived by Wheeler quoted earlier in this chapter, **10⁹⁴ g/cm³**, was calculated using the Plank Length as the lower limit for wavelength. Alternatively the energy and frequency at that wavelength can be thought of as the upper limit for energy and frequency. Since the total energy of the vacuum is then non-infinite we can at least consider non-infinite differences in force strength. Because all matter is not crushed, however, we know that most force interactions must be limited at much lower energies than that of the Planck Length. We can compute this energy density ρ ourselves using Equation 2-6.[25] The Plank length has to be converted to an angular frequency by equating it to a regular frequency and multiply it by 2π, $\omega = 2\pi l_p$.

Equation 2-6

$$\rho = \frac{\hbar\omega^4}{8\pi^2 c^3}$$

The solution is an energy density of **5.7 x 10[118] GeV/cm³**, which is equivalent to the mass energy in **10[95] g/cm³**. Note that **1 GeV** is roughly the mass-energy of a single proton. This result is within one order of magnitude of Wheeler's estimate. Even so, we have no idea if there even is a cutoff energy. And whether there is a cutoff energy or not, the highest energy vacuum fluctuations have to through some mechanism, not be involved in the day-to-day force interactions we measure. The next chapter will give us a guide as to how we avoid thc infinity problem.

The universe we see, feel, and measure is nothing but uncommon events involving insignificant amounts of energy. The real important stuff happens on the zero-point energy scale, not our own. Past physics models that treat the energy of the vacuum as insignificant are incorrect. They have, if anything an inside-out perspective. Instead of seeing how the relatively minor parts of the universe we care about function with respect to the vast energy of the vacuum, physicists have chosen to ignore the energy of the vacuum and treat the parts that matter to us as if they were large in energy and the vacuum insignificant. In order to truly understand our universe we are going to have to see it from the vantage point of zero-point energy, not from our own matter-centered vantage point.

In each chapter as we proceed there will be key points about the zero-point universe that will be useful to keep track of. Many will be significant changes in how we view our universe. The key points from this chapter are:

1) The vacuum is filled with zero-point energy
2) The universe must be understood from the perspective of zero-point energy interactions

Note: For the physicist interested in an in depth examination of zero-point energy theory through 1990, I highly recommend Milonni's book *The Quantum Vacuum.*[25]

[11] J. A. Wheeler and C. Misner, Geometrodynamics, Academic Press, New York, 1962

[12] M. Planck, Zur Theorie des Gesetzes der Energieverteilung im Normalspektrum, Verhandl. Deutsch. Phys. Ges, p.237 (1900)

[13] M. Planck, "Über das Gesetz der Energieverteilung in Normalspektrum." Ann. Physik 4, 553, 1901.

[14] M. Planck, "Eine neue Strahlungshypothese," Verhandl. Deutsch. Phys. Ges. 13: 138 (1911)

[15] A. Einstein, L. Hopf "On a theorem of the probability calculus and its application to the theory of radiation". Ann. Phys. 33: 1096–1104 (1910).

[16] A. Einstein, O. Stern, "Einige Argumente fuer die Annahme einer Molekularen Agitation beim absoluten Nullpunkt" Ann. Phys. 40: 551. (1913).

[17] W. Nernst, Über einen Versuch, von quantentheoretischen Betrachtungen zur Annahme stetiger Energie Änderungen zurückzukehrenVerh. Dtsch. Phys. Ges., 4, pp. 83-116 (1916). (translations and useful commentary from P.F. Browne, The Cosmological Views of Nernst: an Appraisal, APEIRON Vol. 2 Nr. 3 July 1995)

[18] H. A. Lorentz, The Theory of Electrons, 2nd Edition , B. G. Teubner, Leipzig; G. E. Stechert & Co., New York, 1916.

[19] A. Einstein, "Æther and the Theory of Relativity," Address delivered on May 5th, 1920, at the University of Leyden, Germany

[20] S. N. Bose. "Plancks Gesetz und Lichtquantenhypothese", Zeitschrift für Physik 26:178-181 (1924).

[21] W. Heisenberg: Über den anschaulichen Inhalt der quantentheoretischen Kinematik und Mechanik. In: Zeitschrift für Physik. 43, S. 172–198. (1927)

[22] A.E. Dolbear, Modes of motion; or, Mechanical conceptions of physical phenomena Boston, Lee and Shepard (1897) Pg 64

[23] W.E. Lamb, R. C. Retherford, "Fine Structure of the Hydrogen Atom by a Microwave Method". Physical Review 72: 241–243. (1947).

[24] H. B. G. Casimir, and D. Polder, "The Influence of Retardation on the London-van der Waals Forces", Phys. Rev. 73, 360-372 (1948).

[25] P.W. Milonni, The Quantum Vacuum, Academic Press LTD, London 1994, p. 49

Chapter 3: The Casimir Effect, The Model Force

It is found that the influence of retardation leads to a reduction of the interaction energy by a correction factor which decreases monotonically with increasing distance R. This factor is equal to unity for R small compared with the wave-lengths corresponding to the atomic frequencies, and is proportional to R^{-1} for distances large compared with these wave-lengths.[26]

Hendrick Casimir & Dirk Polder, 1948

The Discovery

The discovery of the Casimir Effect was the breakthrough for the mechanical theory of forces. There are other phenomena attributed to action due to vacuum fluctuations, but none is as compelling a proof of the existence of zero-point energy. The Casimir Effect was first proposed by Hendrick Casimir and Dirk Polder in 1946 and 1948 in their identically titled papers "The Influence of Retardation on the London-van der Waals Forces."[26,27] Their effect is based on the underlying principle that the vacuum fluctuations are induced dipoles and that these dipoles interact with each other and produce pressure. This theory is not just important as a proof of the existence of vacuum fluctuations when it was confirmed experimentally,[28,29] but it is the first general description of how vacuum fluctuations produce forces.

Van der Waals Forces

Van der Waals Forces, named after Johannes Diderik van der Waals, are known to be very important with respect to chemical bonds and electrostatic forces between molecules. These are forces between electric dipoles, meaning they have positive and negative charges on opposite ends of a molecule or atom. In the case of vacuum fluctuations it is the fluctuations themselves that are charge dipoles. In much of physics literature vacuum fluctuations have long been treated as being electrically neutral so Casimir's theory was a major conceptual breakthrough. This will be discussed in more detail latter.

The Dutch physicist Peter Debye perhaps best described van der Waals Forces using the illustration shown in Figure 3-1.[30] In the normal case with static electric dipoles, they can have two opposite charge orientations, I and II, and every possible orientation in between.

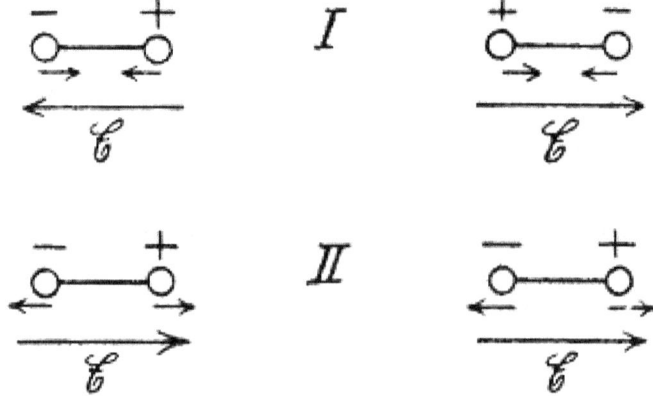

Figure 3-1 Debye's illustration of pairs of dipoles shown in opposite charge orientations with the effect on each dipoles electric moment. An arrow pointing in the direction of the negative charge indicates a decrease in electric moment while an arrow pointing in the direction of a positive charge indicates an increase in electric moment.

If we were to imagine a polar gas, with all possible charge orientations observed from a distance much greater than the spaces between charges, there would be no net electric charge and no van der Waals Force observed. In Debye's analysis the van der Waals Forces are seen when we also factor in the electric moment of the individual dipoles. The dipole moment increases as the space between the charges increases, such that it takes more energy to rotate the dipole. In Case I, as the dipoles are repelled from one another the charges in each dipole are attracted. This secondary effect is attractive and reduces the net amount of repulsion between the dipoles. In case II the dipoles are in a repulsive configuration. In this case the repulsion causes the charges within the dipoles to spread out further, thus increasing their electric moments and increasing the net forces between.

In each case the secondary force is an attractive one. Accept for a few rare cases, which are of great interest to physicists, van der Waals Forces are almost always attractive. Van der Waals Forces come in three types depending on whether the dipoles are permanent or induced. The first type occurs between two permanent dipoles and is called a Keesom Force; the second is between a permanent and induced dipole and is called a Debye Force; and the third occurs between two induced dipoles and is called a London Dispersion Force.[31] Each is named after its discoverer. Cases where dipoles are induced are cases where the molecule or atom is normally electrically neutral, but under the influence of the electric field of another dipole, can become a dipole.

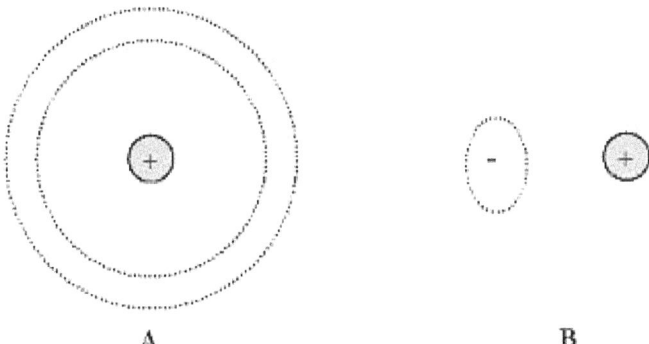

A   B

Figure 3-2 One hydrogen atom **A** where the electron randomly propagates a shell around the proton, such that the atom appears electrically neutral. The second hydrogen atom **B** is an induced dipole with the electron primarily occupying a region off to one side of the proton in response to a local electric field.

One of the simplest examples is a hydrogen atom as illustrated in Figure 3-2, which contains a negatively charged electron in space around a positively charged proton. Under normal circumstances the electron's position is so random that it can be approximated as a shell, making the hydrogen atom appear electrically neutral. In the presence of an electric field, however, the two particles will favor a polar alignment with the field such that opposite charges are adjacent to each other. Under these conditions the hydrogen atom becomes an induced dipole. As the title of Casimir's paper suggests, the forces responsible for the Casimir Effect were thought to be London-van der Waals Forces between induced dipoles, where those induced dipoles are vacuum fluctuations.

Casimir Effect Fundamentals

The second part of the title of Casimir's paper requires that the effect of these van der Waals Forces are retarded, or reduced. This retardation is most easily described using the two parallel plate example. Figure 3-3 illustrates what happens using two plates represented by the two dark rectangles and vacuum fluctuations of various wavelengths represented by ovals of various sizes.

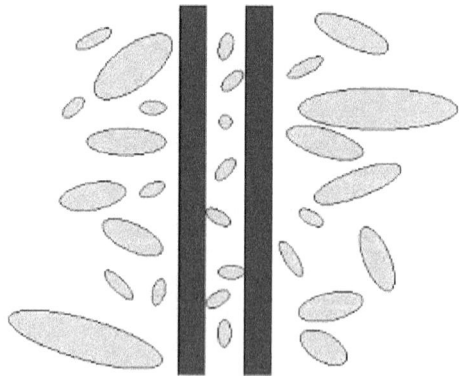

Figure 3-3 A simple illustration of the Casimir Effect between two plates. The force between the two plates is retarded because the longer wavelengths are excluded from the cavity between the plates.

In open space, vacuum fluctuations may be of any size, from the size of the Planck Length at the low end to perhaps the size of the universe of the high end. Van der Waals Forces are, however, limited severely by distance, and the longer wavelength fluctuations are not very energetic, so they are not very important to the Casimir Effect. It is the shorter wavelength, more energetic vacuum fluctuations that produce the greatest amount of pressure. What we find in the parallel plate case is that vacuum fluctuations with wavelengths that are longer than the distance between the plates will be excluded from that space. These longer wavelengths will however still exist outside the plates.

There is a van der Waals derived pressure from the vacuum that pushes on the plates, and that pressure is dependent on the energy of the vacuum fluctuations. By putting two plates in close proximity we have created a condition where the pressure from outside the plates pushing them together is slightly greater than the pressure from between the plates pushing them apart. With most of this energy originating from even smaller vacuum fluctuations than those shown in Figure 3-3, the forces stay pretty close to equal. When the distance between plates is large, the net force pushing on them is negligible.

As the plates are brought closer together, an increasing number of shorter and shorter wavelength, higher and higher energy, zero-point vacuum fluctuations are excluded from the space between. This eventually leads to minute changes in the balance of pressure. The pressure pushing the plates together stays the same, while the pressure pushing the plates apart becomes slightly smaller. This effect is not noticeable at distances that are more normal to our everyday experience, but once you get down to micron distances or less, the change in pressure becomes significant, and the plates are pushed together.

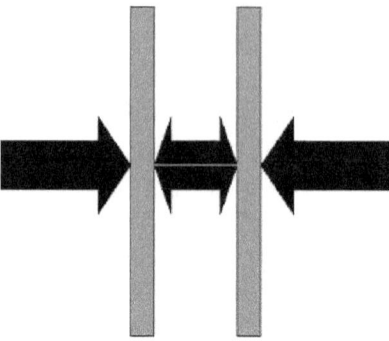

Figure 3-4 A simple graphic illustration of the Casimir Effect between two plates showing how the force represented by black arrows is only slightly reduced in between the plates as designated by the gray line.

Experimental evidence for the Casimir Effect is definitive proof that zero-point vacuum fluctuations are real. It is also proof that they produce pressure that causes bodies to move. More importantly, it provides the best example of how forces work with respect to zero-point energy interactions. The key is that the vast majority of the energy is in the ultra small wavelength vacuum fluctuations and they are pushing on both sides of an object at all times, such that those forces remained balanced. This is simply illustrated in Figure 3-4.

If we had a nearly infinite amount of energy on one side and nothing on the other, the object would be accelerated to the speed of light almost instantaneously. Because these nearly infinite forces instead remain largely in balance, this problem does not exist. We only have to worry about a small pressure differential. Since the force comes about when longer, less energetic wavelengths are excluded, the difference in pressure and thus the magnitude of the force is very small.

Mathematically the Casimir Force between two plates can be expressed by a simple equation, where the force **F** is equal to term consisting of several constants multiplied by the area **A** of the plates and divided by the distance **d** between the plates to the 4th power. Having **d⁴** in the denominator means that the force will be very weak at longer distances, but gets much stronger as the plates are brought closer together, that is, up until the point that molecular interactions become more important.

Equation 3-1

$$F = \frac{\hbar c \pi^2}{240 d^4} A$$

The Casimir Effect is the first experimentally confirmed force theory that illustrates how bodies of matter such as plates move because they are being pushed on by vacuum fluctuations. The ability of the vacuum to push matter is an important principle of force interaction, and as we shall see, is a model for all force interactions that will follow. All forces fundamentally cause pressure differentials such that the pressure exerted by the vacuum fluctuations is different on opposite sides of a body.

The treatment of vacuum fluctuations as dipoles was our first step in treating the vacuum of space as a polarizable medium, which we will see is critical to all forces. The van der Waals Force mechanism is also critical as that is something we will see again and again as the fundamental mechanical principle behind forces due to zero-point vacuum fluctuations. The Casimir Effect is also the first theory that meets the requirements for a completely mechanical theory of force interaction based on zero-point energy interactions. It is little wonder that this small force has attracted so much attention by heretical physicists that oppose the Standard Model and the Standard Model's true believers alike. As we shall see, the introduction of the Casimir Effect was a key turning point in improving our understanding of the zero-point universe.

There are a number of key points from this chapter.

3) Vacuum fluctuations can be treated as induced electric dipoles

4) Van der Waals Forces between vacuum fluctuations produce a pressure from the vacuum, which is exerted on ordinary matter

5) The Casimir Force is caused by zero-point energy

6) The infinity problem with respect to total vacuum fluctuation energy is solved because the vast majority of that energy remains balanced on either side of a body of matter

7) The Casimir Effect illustrates how forces occur with respect to bodies of matter being pushed by vacuum fluctuations

8) The Casimir Force is a model for other force interactions

[26] H.B.G. Casimir, D. Polder, "The Influence of Retardation on the London-van der Waals Forces", Phys. Rev. 73, 360-372 (1948).

[27] H.B.G. Casimir, D. Polder, "The Influence of Retardation on the London-van der Waals Forces", Nature 158, 787-788 (30 November 1946) doi:10.1038/158787a0

[28] S. K. Lamoreaux, (1997). "Demonstration of the Casimir Force in the 0.6 to 6 μm Range". Physical Review Letters 78: 5. doi:10.1103/PhysRevLett.78.5.

[29] U. Mohideen, A. Roy, (1998). "Precision Measurement of the Casimir Force from 0.1 to 0.9 μm". Physical Review Letters 81 (21): 4549. doi:10.1103/PhysRevLett.81.4549.

[30] P. Debye, "Die van der Waalsschen Kohäsionskräfte" Physikalische Zeitschrift, Vol. 21, pages 178-187, 1920.

[31] V.A. Parsegian, Van Der Waals Forces, A Handbook for Biologists, Chemists, Engineers, and Physicists, Cambridge University Press, Cambridge, 2006.

Chapter 4: What is a Virtual Photon?

Positive and negative electrons would rush together and annihilate one another.[32]

Sir James Hopwood Jeans, 1904

In Search of the Vacuum Fluctuations

It has been established that there are vacuum fluctuations, but we need to know what they are. Planck's original theory of the quantum harmonic oscillator was with respect to black body radiation. Black body radiation implies light quanta given off due to heat. Since the temperature of the vacuum has been measured and found to be very low, ~2.7 Kelvin, we know that the vacuum fluctuations are not due to black body radiation, at least not in the usual sense of a body with a certain temperature emitting a spectrum of light related to that temperature. The spectrum of vacuum fluctuations is far too broad and high energy. Instead, the vacuum fluctuations have been simply presumed to be virtual light quanta, literally a quantity of light. The term "virtual" is meant to refer to the short-lived nature of vacuum fluctuations, and is in no way meant to construe that they are not real. Later the term "photon" was coined and was accepted as a substitute for the term "light quanta," leaving us with the common statement that the vacuum fluctuations are virtual photons.

One consequence of vacuum fluctuations being described as virtual photon is that the zero-point field was seen as being electrically neutral. Zero-point vacuum fluctuations were then not considered as being able to interact with matter electro-magnetically. Furthermore, vacuum fluctuations have to exist within the energy-time constraints of Heisenberg's Uncertainty Principle. They have to be undetectable and, conversely, cannot transfer their self-energy directly to matter. Putting those two things together, we get a field of zero-point energy that is not detectable and does not noticeably interact with matter, hence scientists felt free to ignore it. Of course there were others that either denied Nernst's theory of vacuum fluctuations, or simply viewed it as just another æther theory that they could ignore as unnecessary regardless of whether or not it was fundamentally correct.

While there were historical speculations on positive electrons and antimatter, such as in the James Jeans quote above, they are generally not deemed to be consistent with modern theory. It is thought that Jeans, for example, was actually speculating about the proton. Nonetheless, the fact that someone would come up with an idea so far in advance of the mathematical theory and physical evidence does reflect on the underlying elegance of the principle behind the matter-antimatter relationship.

Dirac

The modern theory of positive electrons and antimatter begins with Dirac in 1928.[33] That is when he published his famous equation on the electron and it held an additional solution for what was to become known as the positron. It was also recognized fairly rapidly that the electron and positron could annihilate each other yielding energy.

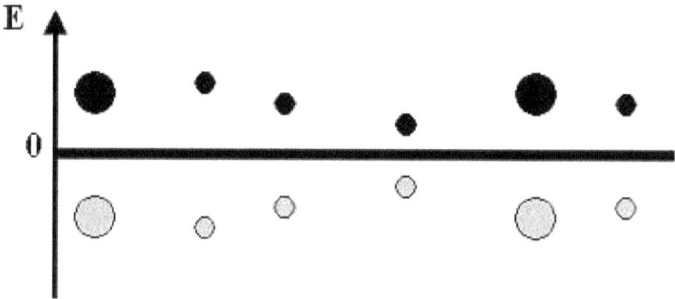

Figure 4-1 An illustration of the Dirac Sea where pairs of virtual particles exist on opposite sides of a line indicating zero energy. The black circles are positive and the gray are negative. We can also think of the 0 line being the division between matter and antimatter.

Shortly after that, Dirac began to speculate about the vacuum being filled with electrons and positrons in what is called a Dirac Sea as illustrated in Figure 4-1. He saw the positrons as holes in space. They can alternatively be viewed as anti-particles in positive space, since we do not have a reasonable model for a hole in space. As of this time, we also do not have knowledge of the makeup of either electrons or positrons, so we should feel free to visualize them however we may wish for the time being.

Carl Anderson discovered the positron in 1932.[34] At that point Dirac's mathematical theory of the positron became accepted as fact. What is interesting is that his theory of the Dirac Sea was not the subject of much scientific development. A Dirac Sea model with space filled with virtual electron-positron pairs gives us a very different model for zero-point energy. Instead of virtual photons, we now have electrons and positrons that are produced in pairs from out of nothing and then rush back together and annihilate, returning to nothing, all within the energy-time parameters of Planck and Heisenberg.

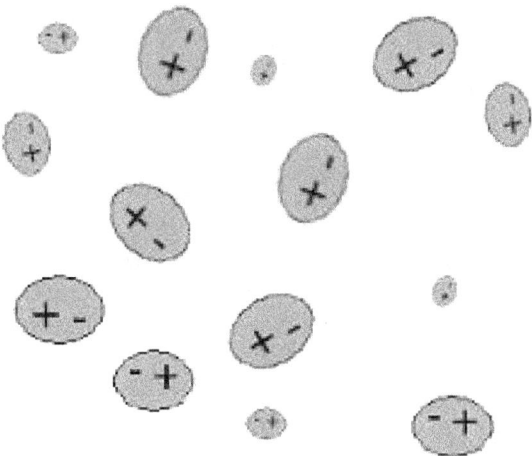

Figure 4-2 A different illustration of the Dirac Sea where virtual particle pairs are shown as oval shaped electric dipoles filling space.

Now we see a vacuum filled with electric dipoles that can orient with respect to charge and rotate in response to charge motion as illustrated in Figure 4-2. Instead of having a zero-point field that is not seen to interact with matter, the Dirac Sea absolutely must interact with matter electromagnetically, at least with anything with non-zero electric charge.

With the concept of antimatter firmly established it was soon extended to photons. A photon is considered to be its own antimatter. The view of vacuum fluctuations in this scheme is that they are virtual photon-antiphoton pairs. Once again the perception based on this model is that vacuum fluctuations are electrically neutral and do not interact electrically. Most physicists up until more recently, with developments in quantum field theory, continued to ignore vacuum fluctuations as being non-existent or irrelevant. Much of zero-point field theory outside of the Casimir Effect has suffered under this ideology. The Casimir effect as presented in the last chapter, however, depends on London-van der Waals Forces due to a polarizable vacuum medium, which favors the Dirac Sea type model. Still much of the physics community apparently thinks of vacuum fluctuations as only having some kind of secondary induced dipole character rather than being true electric dipoles.

Given these two choices of the photon-antiphoton or electron-positron model, which one is correct? Some physicists choose to ignore the Casimir Effect and stick with the virtual photon model. Some consider the polarity as a secondary effect related to the electric and magnetic fields of the photon. Some refer to both models depending on which is most convenient when addressing a given problem. Few, if any, chose the polarizable vacuum fluctuation only model. If you looked up "polarizable vacuum" on Wikipedia the day this paragraph was first written, you would see that it begins "In theoretical physics, particularly fringe physics, polarizable vacuum ... ."[35] That is what polarizable vacuum is to many Standard Model theorists, "fringe physics." Even with a foundation set by Dirac and physical evidence of the Casimir Effect, it is generally thought to be science fiction by the mainstream physics community. As we will see, the overall lack of respect for the polarizable vacuum model has a lot to do with the stalling of fundamental physics over the past eight decades. This lack of acceptance of the polarizable nature of the vacuum is due to the erroneous idea that vacuum fluctuations are virtual photons.

The Photon Mistake

To see why it is an error, we only need to start with the supposition that vacuum fluctuations are pairs of photons and see where that leads us. First we may imagine that a pair of photons consists of two photons being produced out of the vacuum, starting at some central point, then going out a wavelength, and returning another wavelength to finally annihilate each other back in the center. That would give us something like what is illustrated in Figure 4-3. Note that **λ** (lambda) is the symbol for the wavelength.

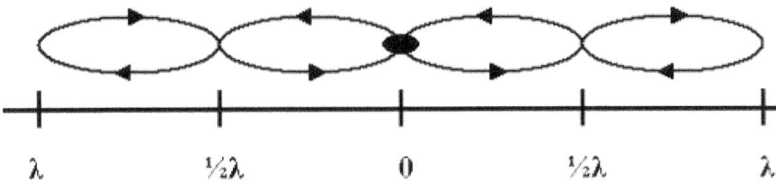

Figure 4-3 A vacuum fluctuation as a photon pair originating at the center with each photon traveling out one full wavelength and then back one full wavelength. The arrows show the direction of propagation.

Now we need to see if this representation meets the energy-time constraints for a vacuum fluctuation. The energy-time constraint due to Heisenberg was discussed previously and is shown again in Equation 4-1.

Equation 4-1

$$\Delta E \Delta t = \frac{h}{2}$$

We can compare that energy-time condition to the energy of a photon as expressed by the well known relationships in Equation 4-2 where **h** is Planck's constant, **c** is the speed of light, **ν** is the frequency and **λ** is the wavelength and the subscript **ph** indicates that this is the photon frequency and wavelength.

Equation 4-2

$$E = h\nu_{ph} = \frac{hc}{\lambda_{ph}}$$

Equation 4-2 expressed in terms of energy-time for a single wavelength would be as shown in Equation 4-3.

Equation 4-3

$$Et = h$$

Looking back at Figure 4-3 we can determine the total energy-time of the photon pair in the illustration. Since each photon goes out one wavelength and back one wavelength, the total number of wavelengths is four, and the total energy-time is **4h**. That amount of energy is eight times the amount of a Plankian quantum harmonic oscillator and eight times the amount allowed under the Uncertainty Principle, so the proposed model in Figure 4-3 cannot be correct.

Next we can consider a photon pair where each photon goes out a half wavelength and returns as illustrated in Figure 4-4.

42

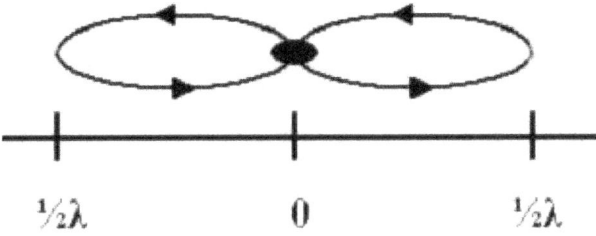

Figure 4-4 A vacuum fluctuation as a photon pair with each photon traveling out a half wavelength from the center and then traveling back. The arrows show the direction of propagation.

At a glance that looks more like what we would expect a vacuum fluctuation to look like. The problem is that there are still four half wavelengths, or two full wavelengths for a total energy-time of **2h**, which is still four times the amount allowed by the Uncertainty Principle.

To continue, we can consider a photon pair of a quarter wavelength each as illustrated in Figure 4-5.

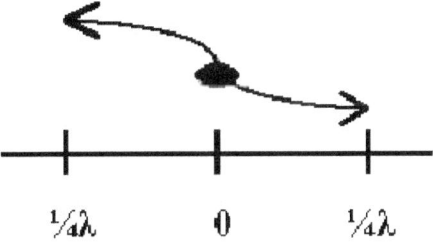

Figure 4-5 A vacuum fluctuation as a photon pair with each photon traveling out a quarter wavelength. The arrows show the direction of propagation. Note that the return path in this case exactly follows the outward path.

This situation is more difficult to imagine as now we have two photons being deflected back along their own paths effectively canceling themselves. Quarter-wave filters are used in optics to filter out specific wavelengths, so similarly such a situation cannot exist, as both photons would be cancelled. In addition to that issue, there is still twice the energy allowed by the Uncertainty Principle and twice the energy allowed for a quantum harmonic oscillator at the lowest allowable energy.

Next we can attempt to reduce the wavelength to an eighth as shown in Figure 4-6. Finally we have the correct energy-time, but there is no viable description of an eighth of a wavelength photon pair. If a wave is deflected at one-eighth its wavelength, it does not return to the starting point, so it is not possible for the two photons to recombine and annihilate. If it could return along its own path, each photon would be cancelled out. More generally the one-eighth photon wavelength photon pair model makes no sense physically. This model is certainly not correct, yet it was our only possibility for a virtual photon pair to have the correct energy. Zero-point energy is not composed of virtual photon pairs.

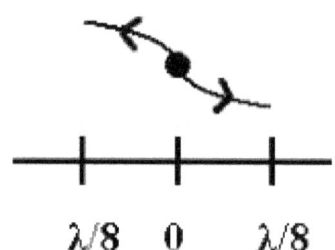

$\lambda/8$ 0 $\lambda/8$

Figure 4-6 A vacuum fluctuation as a photon pair with each photon traveling out an eighth wavelength from the center. The arrows show the direction of propagation. There is no viable return path in this case.

Virtual photon pairs cannot exist within the energy-time constraints required by the theories of Planck and Heisenberg. In order to get the energy-time correct we are limited to only a single half wavelength of a single photon as shown in Figure 4-7.

44

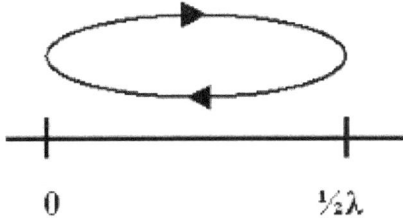

Figure 4-7 A vacuum fluctuation as a single photon of a half wavelength. The arrows show the direction of propagation.

If we only have a single photon, there is no way for it to be produced out of a vacuum, as there is no antimatter photon. Without the photon being initiated as a matter-antimatter pair, there is no way for the photon to be annihilated and the energy returned to the vacuum, or even be produced from the vacuum in the first place. Zero-point energy cannot be composed of single virtual photons either. To find the solution we have to change our approach and look for a more fundamental particle pair than a photon. But what can be more fundamental than a photon? Not surprisingly, that leads us back to the electron-positron pair.

The Fundamental Description of a Photon
There is no fundamental model for the photon that is accepted by physicists. Some would say that there is no known fundamental model for the photon, but that would be incorrect. Physicists have known since the 1930s that a photon during a half-wavelength period can be thought of as a virtual electron-positron pair that rotates 180 degrees. They have simply chosen not to apply that knowledge outside of quantum electrodynamics. A rotating virtual electron-positron pair produces an electric field in its plane of rotation and a magnetic field in the perpendicular direction. The electric and magnetic field is precisely equivalent to the known electric and magnetic fields of a photon during a half wavelength. The curious thing is that while this has been known for most of a century, this photon model is

seldom if ever mentioned as possibly being the fundamental description of a photon.

The electron-positron model of the photon saw serious development at the hands of Richard Feynman and is now commonly seen in a Feynman diagram representation as shown in Figure 4-8.[36] In the figure, the photon is traveling from left to right represented by a wavy line, and an electron-positron pair is represented by the oval. The electron is moving in the same direction as the photon, while the positron is moving in the opposite direction.

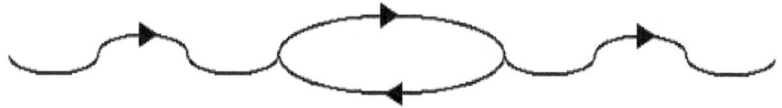

Figure 4-8 A virtual electron-positron pair as part of a photon.

This representation is widely accepted, and it is recognized that the polarizability of the vacuum that leads to the Casimir Effect and other zero-point energy effects is due to the electron-positron pair. But, for some unknown reasons, the mainstream physics community has failed to recognize that the electron-positron model of the photon is the fundamental description of the photon. From here on the electron-positron model is the fundamental model of the photon, and as we shall see it allows us to describe what happens in our universe in a much simpler, more intuitive fashion.

To continue with the development we can visualize a single rotating electron-positron pair as shown in Figure 4.9 with – and + signs indicating the electron and positron respectively, the curved arrow representing the direction of rotation, and the electric (**E**) and magnetic (**B**) fields directed as shown.

E ↑

+
↻
–

B ⊙

Figure 4-9 As a virtual electron-positron pair rotates it produces an electric and magnetic field. Here the pair is shown rotating clockwise with the electric field pointing upward and the magnetic field pointing into the page.

Now we have a virtual particle pair composed of one matter particle and one antimatter particle, one with negative charge and one with positive charge. By definition a virtual electron-positron pair can only exist at the energy of a quantum harmonic oscillator while at the same time meeting the energy-time limits of the uncertainty principle. It is important to note, and will be repeated here numerous times, that these virtual pairs exist over a continuum of energies and are not limited by the rest mass energy of their stable electron and positron counterparts. Indeed they should be considered as something quite different, so think of the names electron and positron here as a mere convenience, which allows us to identify the electric charge and matter-antimatter properties.

Perhaps you are not convinced about the energy equivalence of a virtual electron-positron pair to a half wavelength of a photon. We can look at the energy relationship as it applies to a photon by either frequency υ or wavelength λ as previously shown in Equation 4-2. In order to be equivalent to a photon over half the photon wavelength, the wavelength of the virtual electron-positron pair must be half the photon wavelength. The wavelength of the electron-positron pair is designated by subscript z and shown in Equation 4-4.

Equation 4-4

$$\lambda_z = \frac{\lambda_{ph}}{2}$$

Then we can consider that the maximum energy of a virtual electron-positron pair is the same as the well known expression for the maximum energy allowed as defined by Heisenberg's Uncertainty Principle. This is Equation 4-5.

Equation 4-5

$$E = \frac{h\nu_z}{2}$$

Now we convert to a wavelength equivalent equation and substitute into the prior wavelength relationship we get the following Equation 4-6.

Equation 4-6

$$\frac{h\nu_z}{2} = \frac{hc}{2\lambda_z} = \frac{hc}{\lambda_{ph}}$$

This shows that an electron-positron pair carries the same energy as a photon during a half wavelength. This also proves the assertion related to Figure 4-5 that if we had two virtual electron-positron pairs existing at the same time as part of a single zero-point event, it would violate Heisenberg's Uncertainty Principle by existing too long to be undetectable.

Conclusion

Getting back to our question about whether the zero-point field is filled with virtual photon-antiphoton pairs or virtual electron-positron pairs, we can now say with absolute certainty that of the two choices it can only be virtual electron-positron pairs. If you start with the idea that vacuum fluctuations are virtual photons, they end up having to be a virtual electron-positron pairs anyway. We will see that this idea of zero-point energy being composed of oppositely charged matter-antimatter pairs can be extended to include any other real matter-antimatter particle pair that can form an electric charge dipole. Virtual proton-antiproton pairs are one example, although it may be better to think of them as a particle pair with proton-like matter and electric charge orientation. Proton-antiproton particles pairs will turn out to be critical to understanding forces and will be discussed at length in coming chapters. In general we can state that zero-point energy is composed of virtual matter-antimatter particle pairs.

In summary, we can say that this chapter has given us the following points:

9) Virtual photons do not exist
10) In theories, virtual photons must be replaced by virtual particle pairs with charge
11) The zero-point field consists of virtual particle pairs with charge
12) The vacuum is polarizable

Note that much of the above information was first published in my book *The New Physics* in 2001.[37]

[32] J.H. Jeans, "A suggested Explanation of Radio-activity" Nature 70 pg 101 (June 2, 1904)

Also see JEANS, '"The mechanism of radiation". Phil Mag (6) 2 421-455 (issue 11 Nov. 1901) (ref. 39), p. 426. 42 C. V. BURTON and J H Jeans 1899 Proc. Phys. Soc. London 17 754 Issue 1pp 754-793

[33] P.A.M. Dirac "The Quantum Theory of the Electron", Proc. R. Soc. A (1928) vol. 117, no 778, 610-624

[34] C.D. Anderson "The Positive Electron", Physical Review 43: 491 (1933)

[35] http://en.wikipedia.org/wiki/Polarizable_vacuum

[36] P.W. Milonni, The Quantum Vacuum, Academic Press LTD, London 1994, p. 49

[37] R Fleming, The New Physics, self-published, 2001 (rayfleming.com)

Chapter 5: The Zero-Point Photon

We shall first choose the special, simple case that all oscillators are in their ground states initially and finally, so that all photons are virtual.[38]

Richard Feynman, 1950

The Virtual Photon

Feynman recognized over 60 years ago that all photons could be thought of as being virtual in nature. His conclusion was reached based on the points that a photon is only detectable at the times it is emitted and absorbed. In between those times it is effectively virtual as we cannot detect it or determine its form directly. To be clear, he was using the word "virtual" in a different sense than previously used here. These photons have energy that was given to them when they were produced, and that energy propagates through space without violating the Uncertainty Principle. Between this recognition and the electron-positron model of the photon, Feynman perhaps came the closest to a real breakthrough in understanding the fundamental nature of the photon. After all, if we cannot know the nature of the photon wavelengths directly, other than through their electric and magnetic fields and states of polarization, the electron-positron model makes more sense. He should have realized it, as should every physicist who followed.

Once we understand that virtual photons do not exist and must instead be virtual particle pairs, we can extend that principle to photons in general. In the virtual particle pair model of the photon, a photon is nothing more than a series of virtual particle pairs. Each particle pair in the series serves to conserve energy and momentum while neutralizing the previous electric and magnetic field. A visual representation of a photon is shown in Figure 5-1.

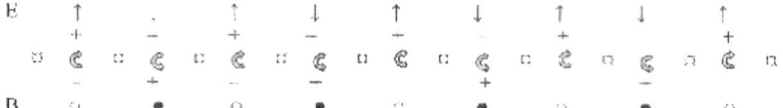

Figure 5-1 A series of alternately rotating virtual electron-positron pairs produces an alternating E field (up and down) and an alternating B field (in and out) identical to that of a photon.

Understanding how energy propagates from particle pair to particle pair is straightforward. We can consider a simple coil of wire to illustrate. As current moves through the wire it generates a magnetic field. If the current in the coil is cut, a back Electromotive Force (EMF) is generated causing a current in the reverse direction. If there is enough energy in the coil it may produce an arc to ground. If not, it may oscillate, producing heat in the coil in order to dissipate the energy. The magnetic field is reversed after the current in the coil is cut, and then the reversed magnetic field induces a back EMF in the coil.

The situation with a photon is similar. The first particle pair in a series produces an electric and magnetic field. When that pair is annihilated, the surrounding field reverses electric and magnetic polarity in an effort to neutralize the previous field. Those fields induce a new central particle pair to form with the appropriate electric and magnetic properties. The new particle pair is also induced in such a way that the energy and momentum of the photon is conserved. In this way it is produced out of vacuum in exactly the correct position, with the correct energy and state of polarization. Each successive particle pair in the photon chain forms because of this electro-magnetic induction effect. Even more interestingly, this is true in the general case, virtual particle pairs will always be induced into existence such that energy, momentum, and electric and magnetic field properties are conserved.

To complete our conceptual analysis of a photon we need to recognize that the electric and magnetic fields induced by the rotating series of central virtual particle pairs are also virtual particle pairs. A photon should be thought of as a collection of virtual particle pairs with a series of central pairs and the others spread across the zero-point field as simply illustrated in Figure 5-2. It is the associated virtual particle pairs across the zero-point field that form a wave front and give a photon its wavelike properties as they oscillate in response to the central pair's motion and electro-magnetic orientation.

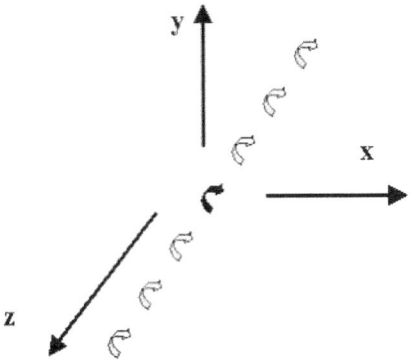

Figure 5-2 A photon moving in the x direction with a central pair (black) surrounded by a field of particle pairs all rotating in response to the central pair's rotation.

In the illustration, the virtual particle pairs are shown rotating with positive and negative charges oriented along the y-axis. The photon's field of virtual particle pairs is only shown along the z-axis but is in reality throughout all space. It is also important to note that the field around the central pair is not composed of particle pairs rotating in exactly the same manner and with the same orientation as the central pair, but are rather somewhat randomized and only slightly biased in the direction consistent with the photon's central pair. The zero-point field is far too energetic for all the vacuum fluctuations to be involved, so only a small amount of change by a small percentage of vacuum fluctuations is required for a photon to propagate through space.

The wave-particle duel nature of the photon is not as complicated as it has long been made out to be. The energy of the photon flows through the vacuum in response to the movement of a series of central virtual particle pairs. The electric and magnetic fields propagate and collapse in time with the central virtual particle pairs. Anything in the way of the wave front can affect the movement of the central pair. When the central virtual particle pair interacts with something it generally acts like a point particle, a physical photon. Such is the case when a photon is emitted or absorbed, with the energy being funneled from or into the central point of interaction. When the wave front of associated particle pairs interacts with something, the photon behaves in a wavelike manner.

All of the known properties of the photon can be explained within the scope of the virtual particle pair model. One of the simpler examples is polarization. In the case of linear polarization, the particle pairs are rotating parallel to the axis of polarization. With a circularly polarized beam of photons the particle pairs are rotating in a circle perpendicular to the direction of propagation in addition to rotating forward. The series of particles in the circularly polarized case ascribe a helical path every half wavelength, a double helix when both charges are considered.

Another point of discussion relates to the rest mass of the electron and positron. Pair production of these two particles requires 1.022 MeV of energy, as they each have a rest mass of 0.511 MeV. Since photons display a vast range of energies, we must be clear that virtual electron-positron pairs are not the same thing as a stable electron and positron. As mentioned previously, virtual particle pairs consist of a continuum of energies both below and above the pair production energies of their stable equivalents. We can also consider that photons can be made of other types of particle pairs. One such possibility is proton-antiproton pairs, which have a pair production energy of 1877 MeV.

The Zepton

In order to avoid confusion between free electrons and positrons or other particles pairs and the continuum of virtual particle pairs, it is best to use another name for virtual particle pairs. In my previous book I used the term "parton," as that has appeared in some literature in a related usage. However, the word "parton" was favored by Feynman as the name for particles that make up the proton.[39] It was later co-opted by quark theorists, including eventually Feynman himself, who took it for their own even though Feynman's original theory referred to something entirely different. Feynman's original theory of the proton deserves future reconsideration so his term "parton" should be reserved for those discussions, keeping in mind that his original partons may turn out to be virtual particle pairs.

The word for virtual particle pairs we will use throughout the remainder of the book is zeptons. This comes from the prefix for units of measure "zepto" (10^{-21}), referring to very small and very energetic particles. It also is a reminder that they are a form of zero-point energy. Keep in mind that the word zepton does not in any way imply that these virtual particle pairs are anything new; it is just to get past some people's difficulty in comprehending the differences between virtual particle pairs and their stable or meta-stable counterparts.

Pair production is easier to understand within the scope of a zepton model of a photon. In a zepton, an electron and positron can already be present and so they do not have to be produced by some secondary effect. The particles are there all along. Given a photon that happens to have ≥1.022 MeV of energy, it will have enough energy to produce a stable 0.511 MeV electron and a stable 0.511 MeV positron. Note that the photon does have to interact with other matter to conserve momentum, so pair production does not occur in free space. Pair production is a simple state transition that occurs when the photon has enough energy for its constituent particles to be free and stable prior to interacting with matter to initiate the pair production process. The same thing occurs when a photon is composed of a proton-antiproton pair and it has ≥1877 MeV of energy. The proton and antiproton are already present.

Considering the broader consequences of this theory of the photon we must recognize that a photon is not a fundamental particle at all. It is the zeptons, the virtual particle pairs that are fundamental, and the zepton model of the photon is a fundamental description of the photon. The importance of the photon then is not as a fundamental particle but rather as an elementary form of energy transport through the zero-point field consisting entirely of zero-point vacuum fluctuations.

In summary, the determinations made in this chapter have given us the following key points:

13) Photons are a series of zeptons with an associated field of zeptons
14) Zeptons are fundamental
15) Photons are not a fundamental particle
16) Pair production is a simple energy transition from virtual to stable or meta-stable particles
17) The zero-point field consists of zeptons
18) Photons are a means of transporting energy through the zero-point field
19) Zeptons are induced into existence with properties that conserve energy and momentum
20) Particle-wave duality of a photon is due to the zeptons of the zero-point field around the photon responding to the central zepton's charge orientation, rotation, and momentum

[38] R.P. Feynman, Mathematical Formulation of the Quantum Theory of Electromagnetic
Interaction, Phys Rev 80:440-457 1950.
[39] R.P. Feynman, "The Behavior of Hadron Collisions at Extreme Energies". High Energy Collisions: Third International Conference at Stony Brook, N.Y.. Gordon & Breach. pp. 237–249 1969. ISBN 978-0677139500.

Chapter 6: Zeptons and Faraday Field Lines

This leads us to a picture of discrete Faraday lines of force, each associated with a charge, -e or +e. There is a direction attached to each line, so that the ends of a line that has two ends are not the same, and there is a charge of +e at one end and –e at the other.[40]
Paul Dirac, 1963

The Mechanism Behind Electricity and Magnetism
Next we need to ask, how does electricity and magnetism work? The biggest problem with the basic electromagnetic theory on a large scale is not with it producing wrong results, but the absence of a technically sound underlying force mechanism that is responsible for the interactions. In currently accepted theory, photons are considered to be gauge bosons, the force carriers for the electromagnetic force, but that is a far from satisfactory answer as pointed out previously. In the theories of the so-called Standard Model, bodies or particles of matter exchange photons in order to figure out how they are to respond electromagnetically. The first difficulty is that these cannot be stable photons, as that would require that bodies give up energy, and they would quickly be reduced to nothing; these gauge boson photons must be virtual photons. But, as we have already seen, virtual photons do not exist.

Next we have the thorny problem of where do these smart gauge photons store their information? If you look at the properties of photons, nowhere is there a memory chip that can store a few dozen bits of information. Even the first bit, say attraction versus repulsion, cannot be explained by a fundamental property of a virtual photon. What about whether it is electric or magnetic? If we consider the size of objects of the universe from say the size of a bare electron, $\sim 10^{-22}$ meters, to the size of the visible universe, $\sim 10^{26}$ meters, we see that the amount of data that a photon needs to store in order to transfer information over that entire range of electromagnetic force dynamics is unfathomably huge. Photons do not have even a single bit of memory to hold data that could tell a body of matter if an electric force is attractive or repulsive. So how can they hold a universe worth of data? The bottom line is that the photon as the gauge boson for electricity and magnetism theory is utter garbage.

To discover the way that electricity and magnetism really works we can begin by looking at what happens with a charged particle, say a stable electron, in free space. If the electron is near other charged particles or polar molecules, they orient themselves with respect to charge, such that positive charges and the positive ends of polar molecules will face toward the electron and its negative charge. If we consider a volume of space filled with hydrogen atoms containing one proton and one electron, they are induced to form dipoles. The electrons and protons in the hydrogen become slightly separate in nearby space with the positive dipole, the proton, preferentially facing toward the local free electron, as simply illustrated in Figure 6-1. In this manner, a free electric charge affects the polarity of nearby polarizable matter.

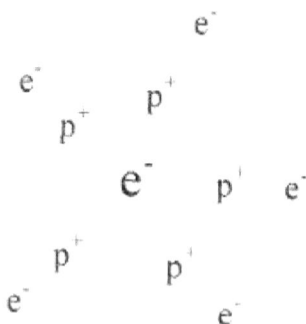

Figure 6-1 A free electron surrounded by hydrogen atoms that are polarized in response to its electric charge.

What do we expect to happen if nearby space is a vacuum? The local vacuum is filled with zero-point energy, with each zepton flashing into and out of existence very quickly. Each vacuum fluctuation that forms the zero-point field is a miniature, short-lived electric dipole. These dipoles orient themselves with respect to nearby electric charges, or at least some of them will. This polarizable zero-point field acts very similarly to the hydrogen gas we considered above.

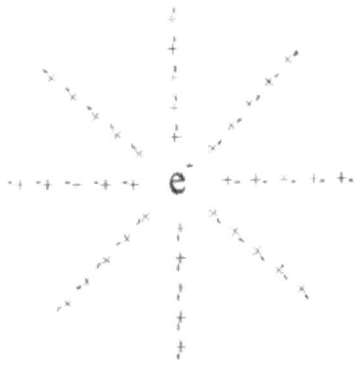

Figure 6-2 A free electron surrounded by zeptons that are polarized in response to its electric charge.

While scientists may treat the vacuum as being nothing, without electric properties, it is in reality a sea of dipoles, a variation on Dirac's Sea. So what happens is something like what is illustrated in Figure 6-2.

Zepton dipoles align with their positive charge, such as a virtual positron, facing toward the free electron and their negative charge, such as a virtual electron, facing away from the free electron. Other zeptons then align themselves with the electron and other nearby zeptons. While Figure 6-2 only shows a few lines, there are polarized zeptons all the way around the charge at every distance, extending toward infinity.

The Nature of Fields
To back up a bit, it may be constructive to review some historical views on electricity and magnetism. Unlike

most simple mechanical devices which operate when one part of a mechanism is in contact with another part and mechanically pushes it, electricity and magnetism was observed to work over large distances without there appearing to be anything in between.

> *The explanation of the cause of the inequality of pressure at once suggests the means of expressing the dipolar character of the line of force. Every vortex is essentially dipolar, the two extremities of its axis being distinguished by the direction of its revolution as observed from those points.*[41]
>
> James Clerk Maxwell, 1861

While not knowing what actually fills the vacuum in between, there was a strong feeling that it needed to be represented by something. An early experimenter in electricity, Michael Faraday came up with the concept of a field to represent this unknown physical mechanism.

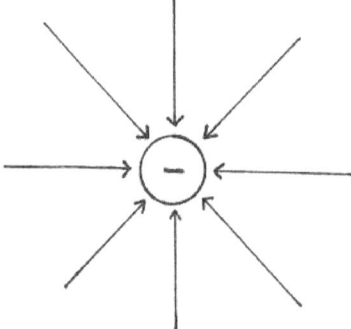

Figure 6-3 A free electron with field lines pointing toward it.

In order to illustrate these fields, he drew them as lines, which became known as field lines, or, in deference to the inventor, Faraday field lines. In Figure 6-3 we can see a simple illustration of the field lines around a single negative charge. The convention is that electric field lines point toward negative charges. This selection was completely arbitrary.

The connection between the aligned polarized zeptons and Faraday's field lines is unmistakable. Note, however, that in Faraday's method of illustration the number of lines was proportional to the strength of the force. The force itself is spread throughout space in the same way that zeptons are spread throughout space.

In the case of the electric field around a point charge, the field strength is uniform at any given distance. The force also declines in inverse proportion to the square of the distance. The easiest way to see how this occurs is to recognize that the same total amount of polarization occurs at each distance away from the charge; it is just spread out over increasingly larger volumes of space. This is illustrated in Figure 6-4.

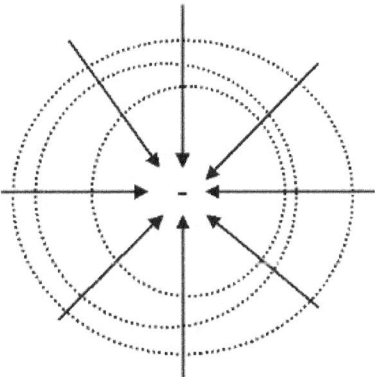

Figure 6-4 A free electron with field lines pointing toward it with different surfaces around the charge with each crossing exactly the same number of field lines.

The area of a sphere is **$4\pi r^2$**, where r is the radius of the sphere. The density of electric charge polarization, the electric field density, decreases with the square of the distance, as the area of the spheres increases with the square of the distance. At very large distances the amount of polarization may become undetectable to our equipment, but in principle it continues to infinity.

In order to further review the Faraday field line representation of electric fields we need to examine the lines that form when there are two opposite charges nearby. Following convention, the arrows are drawn from the positive charge toward the negative charge as shown in Figure 6-5.

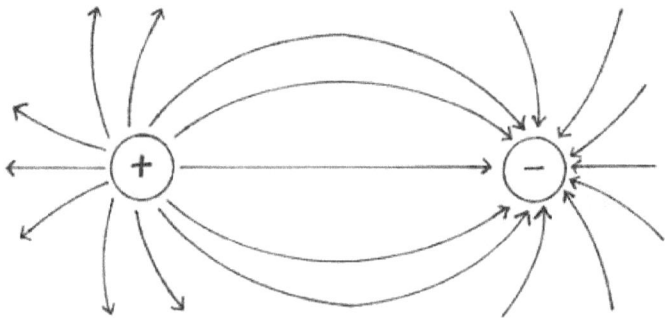

Figure 6-5 A free positive charge and a free negative charge with field lines aligned between them.

Likewise Faraday illustrated electric field lines between like charges roughly as follows in Figure 6-6. This illustration shows the repulsive force between the like charges.

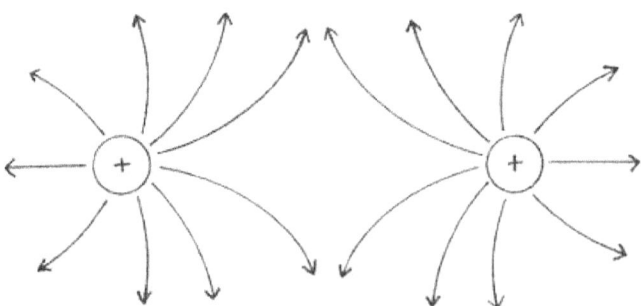

Figure 6-6 Two positive charges with field lines between them.

What is important about the zero-point field model of the vacuum is that there no longer has to be action at a distance. There no longer has to be a mythical field in space being used as a placeholder for something real that we will eventually discover. The thing that is real, the vacuum fluctuations (the zeptons), were discovered over a century ago.

Dirac gave us a hint with his speculation from a half-century ago that perhaps Faraday field lines had something to do with real charges in space, virtual electrons and positrons. Except these virtual electrons and positrons do not just appear at the ends of the field lines but rather the field lines are an illustration of real zepton dipoles that cover the full length of the field lines from one end to the other, or toward infinity. These dipoles are lined up end to end virtually in contact with each other, or to put it another way, in contact with each other, however one defines contact in the virtual particle realm. This is a true mechanical description of electrostatic force interaction.

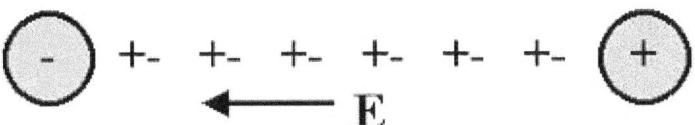

Figure 6-7 The electrostatic or E field points in the direction of the positive end of the zepton dipoles.

Faraday's electrostatic field, frequently designated by the letter E, is simple to convert to zero-point theory since the alignment of the field is in a line with the polarized zepton charges going from negative toward positive as shown in Figure 6-7, keeping in mind that the positive end of the zepton is pointing toward a negative free charge.

Electric Attraction and Repulsion
Positively charged particles partially polarize the vacuum fluctuations in the opposite orientation from negatively charged particles. The word "partial" is used

in an intentional way, because the number of particle pairs that exist instantaneously is very large, and a single unit charge is very small. A small amount of charge only causes a small percentage of zeptons to orient their dipoles in the charge's direction, and most are only oriented to a slight degree.

e- +- +- +- +- +- +- +- +- +- +- +- p+

Figure 6-8 A free electron and proton with a row of polarized zeptons lined up between them end to end across the entire distance.

Returning to the illustrative examples, we can examine more closely how this mechanical force interaction works. What happens now when two particles of opposite charge, say an electron and a proton, are near each other in space? In this case we have the effect illustrated in Figure 6-8 where zeptons form a virtual bridge between the particles. There will be of course a continuum of such lines with most of them curved in a manner consistent with Figure 6-5.

Looking instead at two like charges in space, say two electrons, we have the situation shown in Figure 6-9.

e- +- +- +- +- +- -+ -+ -+ -+ -+ e-

Figure 6-9 Two free electrons with a row of polarized zeptons lined up between them with like charges in opposition at the midway point.

In this case we have two negative virtual charges in virtual contact with each other. These two negative charges are repelled from one another. These two like charges must then move away from each other, if only slightly, before the zeptons are annihilated. As new zeptons are produced in their place they also align with the stable charges, while at the same time being forced to move in response to other nearby zeptons. Generations of zeptons will move in such a way as to produce a more stable situation. As the like charges of the zepton dipoles deflect, they produce a small hole in space. That hole is filled with another zepton, then another, and another, and another. We end up with a situation as illustrated in Figure 6-10.

Figure 6-10 Two free electrons with polarized zeptons between them showing the deflection in the center and a new zepton coming into existence producing pressure that pushes the electrons apart.

As each new zepton arrives on the scene it is forced to align with nearby polarized particle pairs and subsequently deflects to avoid like charges. This creates an increase in pressure in the space between the free particles. The virtual particle pairs push against adjacent pairs, which in turn push against those in line toward the particles. This pushes the stable particles apart as the pressure of the vacuum between the particles increases and overcomes the pressure of the vacuum outside the particles pushing them together. This is the mechanical basis for electrostatic repulsion, and as we can see, the basic mechanism is reminiscent of the Casimir Force as it is dependent on vacuum pressure differential.

As before, there is a continuum of lines of dipoles surrounding each charge. Like the lines of zepton dipoles in the middle, they will all be deflected away slightly. The lines of zepton diploes end up tracing out the well-known Faraday electrostatic field lines previously shown in Figure 6-6. Figure 6-11 shows, the same thing with virtual dipoles. These are not field lines at all, but rather zeptons acting as simple dipoles, instantaneously lined up end-to-end.

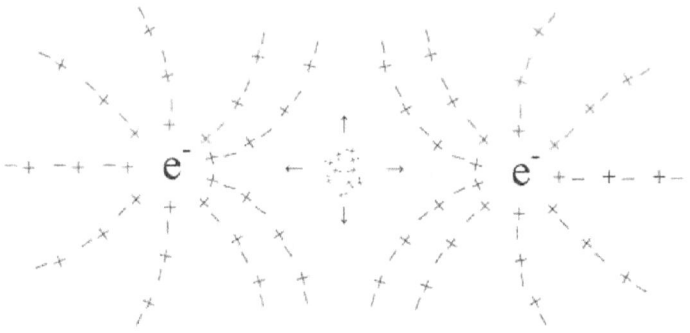

Figure 6-11 Two free electrons with polarized zeptons oriented between them forming the well-known electrostatic repulsion field lines.

With the repulsive case it is somewhat easier to understand how the vacuum pressure between two like charges increases. Returning to the electrostatic attraction case we can visualize polarized zeptons lined up end-to-end between two opposite charges. The key to understanding the force interaction is understanding what happens when each zepton annihilates and disappears. Remember that the electron-positron dipole starts at a point, moves outward to its maximum extent, its wavelength, and then collapses back to a point. As each zepton contracts newer zeptons next to it easily fill in the space. When each successive zepton vanishes, the space available for the next pair is smaller, so smaller wavelength zeptons take its place or longer wavelength zeptons meet in the middle filling the space once occupied by a zepton. The pressure pushing outward between the charges is slightly reduced as compared to the normal pressure pushing on matter from all directions. Over time, the length of the chain of zeptons is shortened as the pressure from outside pushes the charges together. Figure 6-12 illustrates the effect.

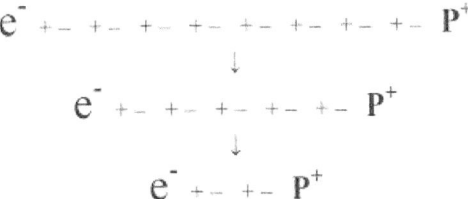

Figure 6-12 A free electron and proton being pushed together as the available space for zeptons between them disappears.

It is important to remember the push principle, the idea that everything moves because it is pushed, as it is tempting to think of the zeptons as pulling on adjacent zeptons. We need to remember that there is a general pressure being exerted on the particles by the vacuum in all directions at all times. We can consider in particular a component of this pressure on the line between the particles. The electrostatic attractive force is not due to the virtual dipoles between the charges pulling them together as we might imagine, but rather the dipole alignment between the oppositely charged particles leads to a reduction in the pressure pushing the stable charged particles apart. As with the Casimir Effect, the pressure pushing the particles apart is less than the pressure pushing them together, so the particles are pushed together.

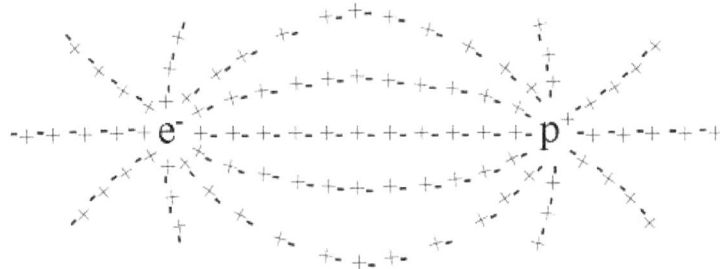

Figure 6-13 A free electron and proton with zeptons between and around them showing the well-known Faraday field lines of electrostatic attraction.

67

As we did before, we can consider the continuum of strands of virtual dipoles that exist all the way around each particle. When we look at the off-axis strands of zeptons, we see that as zeptons are excluded, the strands get pushed toward the axis. What we see then are stands of virtual pairs that follow the well-known Faraday field line model for electrostatic attraction, as seen in Figure 6-13.

Magnetism
Next we can consider the magnetic force. In order to understand the origins of the magnetic field we need to start with a moving charge and consider what effect it has on the virtual dipoles of the zero-point field. To begin, we can consider a free electron moving through space. As with the static case, we see the zepton dipoles orienting themselves with respect to the charge. As before, it is only a small percentage of the total number of zeptons that participate in the interaction, but we shall focus on the few that do. The nature of the zepton dipoles is to align with respect to a charge, so when the free charge is moving, the dipoles change their orientation in order to stay aligned. So, as a free electron moves, the zepton dipoles rotate to maintain their orientation relative to the stable charge as illustrated in Figure 6-14.

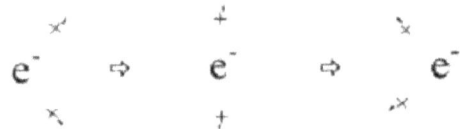

Figure 6-14 A free electron in three separate positions relative to two zeptons. The zeptons rotate in response to the electron motion.

Now we can consider a series of charges in motion instead of a single charge. A series of charges in motion constitutes an electrical current. The zeptons rotate in response to the first charge, and continue rotating as the next charge passes by, and so on, and so on, and the zeptons keep on spinning. Or, more properly as the zeptons flash in and out of existence, each successive one continues the spinning motion, conserving angular momentum.

> *Our result is in fact that a linear current is a vortex ring in the fluid aether, that electric current is represented by vorticity in the medium, and magnetic force by the velocity of the medium. The current being carried by a perfect conductor, the corresponding vortex is (as yet) without a core, i.e., it circulates round a vacuous space.[42]*
>
> Joseph Larmor, 1894

As before, not all zeptons will rotate, as there are a lot more of them than are needed to balance the motion of charge. Each zepton's rotation then affects other adjacent zeptons, causing them to rotate. Then when one zepton dies a new one takes its place in the empty spot conserving angular momentum by continuing the rotation. Figure 6-15 shows an illustration of spinning zeptons next to a series of electrons moving in a straight line such as in a conductor.

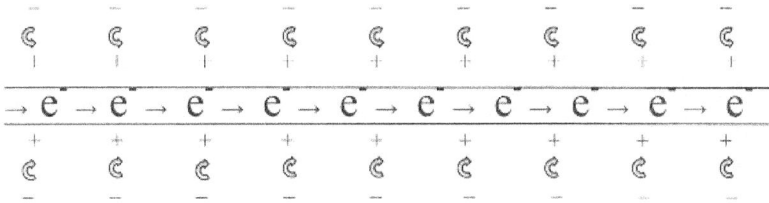

Figure 6-15 Electrons in motion within a conductor induce rotation in surrounding zeptons.

The opposite effect is also true as the sea of spinning zeptons can induce a current in a conductor such as a wire. The angular momentum of the spinning zeptons must be conserved, so not only do replacement zeptons continue spinning, but the current also continues moving. The spin of one zepton induces spinning in other nearby zeptons, which induce replacement zeptons to spin thus continuing the effect. At the same time if there is a nearby conductor, such as a wire, the spinning zeptons induce charge movement in the conductor converting the angular momentum of the spinning zeptons into linear momentum within the conductor. That is how a magnetic field induces current in a conductor.

The balance between the spinning zeptons in the zero-point field and the current in the conductor forms an inherently symbiotic relationship with each relying on the other in order to be sustained. One can think of it as something akin to inertia for electric forces. The current continues as long as the magnetic field continues and the current stops when the magnetic field stops.

> *What is the mechanical cause of this difference in pressure in different locations? We have supposed... that this difference in pressure is caused by molecular vortices, having their axes parallel to the lines of force.*[41]
>
> James Clerk Maxwell, 1861

Faraday's Magnetism

The magnetic field is frequently designated by the letter B, so it is often called a B field for short. To derive the standard magnetic force properties of interaction we can start by referring to the right hand grip rule.

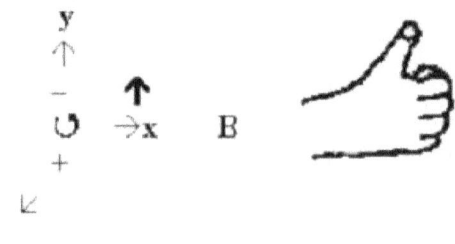

Figure 6-16 The right hand grip rule to determine the direction of the magnetic, or B, field shown here pointing along the y-axis.

If you hold your hand with your fingers curled inward and your thumb pointing up and then align your fingertips so they point in the direction of zepton rotation, your thumb will point in the direction of the B field as illustrated in Figure 6-16. With this simple translational technique it is possible to see how zepton motion and classical magnetic theory relate to one another.

With a common bar magnet the magnetic field lines are shown exiting the north pole looping around the magnet and entering the south pole. A basic illustration of this Faraday's magnetic field line is shown in Figure 6-17 below.

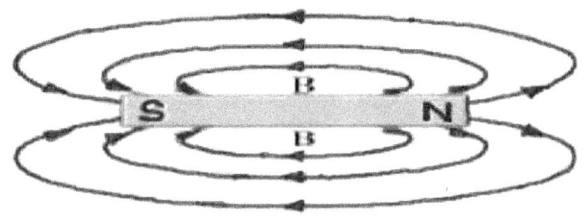

Figure 6-17 A simple bar magnetic with north and south poles marked N and S respectively and magnetic field lines running from one end to the other.

The next step is to tie electric current and magnetism together. The simplest experiment often shown in grade schools is to wrap a wire around a nail made of iron or ordinary steel. Then attach a battery to the ends of the wire. The coil of wire produces a magnetic field that turns the nail into an electromagnet. We can look at the magnetic field that is produced by a simple coil as shown in Figure 6-18.

The right hand grip rule can once again help us determine the direction of the magnetic field produced by the coil. If you curl your fingers and have the fingertips point in the direction that current flows through the coil - in the case above, your fingers are above your palm - then your thumb will point in the direction of the magnetic field, the north pole of the electromagnet.

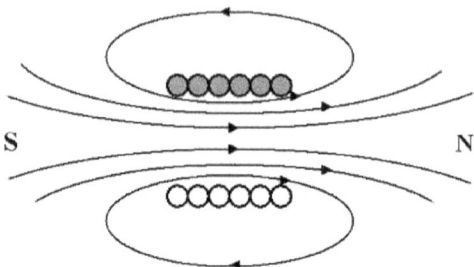

Figure 6-18 A cross section of a coil with a current coming out the top of the winding and in the bottom spiraling from left to right through the windings producing a magnetic field with the north pole to the right.

To string it all together we only need realize that the current in the coil is making the zepton dipoles of the vacuum that are inside the coil spin in the same circular direction. If you imagine those little dipoles as little loops of current, the magnetic field line is like drawing strings through all those little loops so that they lie in a line. It is also important to note that while the zeptons inside the coil spin one way, the zeptons outside the coil spin the opposite direction, and we can see the magnetic field lines outside the coil also point in the opposite direction. This is always the case since if you look back at Figure 6-14 you can see that zeptons on opposite sides of a current carrying wire, always spin in opposite directions.

Permanent Magnets
No discussion of electricity and magnetism would be complete without mentioning ferromagnetism. Ferromagnetism is the ability of some materials, most commonly iron based, hence the use of *ferro* in the name, that can become permanently magnetized. A permanent magnet produces a magnetic field without an external current being present, although a magnetic

field is necessary in order to magnetize the magnet initially. Generally, ferromagnetism is thought to be due to rotating dipoles trapped within the ferromagnetic material. In addition to rotation in the material, there must also be zepton dipoles rotating within the structure of the ferromagnetic material. There is certainly some synergistic effect between the matter of the magnet and the zeptons of the zero point field that sustains the magnetic properties. The exact nature of that effect is yet to be determined.

Magnetic Monopoles

As a bit of a side note we can briefly consider the nature of magnetic monopoles. There are some theories that postulate that magnetic monopoles exist, such that one could have a north pole magnet without a south pole. So far such a thing has never been seen, and standard equations for electricity and magnetism do not allow for them to exist. There is a good reason that one has never been seen. Magnetic fields are rotating dipoles. With a rotating dipole the north pole of the magnetic is in the direction shown by the right hand grip rule in Figure 6-14. The south pole of the magnet can be found by using the same approach with one's left hand. Note that using this technique the right thumb and left thumb will always point in opposite directions with respect to a spinning dipole. Since magnetic forces are due to spinning dipoles they must always produce both magnetic poles, hence there is no such thing as a magnetic monopole. We will also see later that mathematically the magnetic field strength of the north pole always equals that of the south pole so that the net magnetic field around a magnet is always zero. Magnetic monopoles are nothing but science fiction. They are the brainchild of bored mathematicians, but have nothing to do with reality.

Conclusion

Dirac was correct to consider that Faraday Field lines may be composed of virtual charges. He simply did not take the idea to its full and logical conclusion. Starting with the idea that the zero-point universe is filled with virtual dipoles, zeptons, we now have a basic mechanical description of how electromagnetic force interactions propagate through space. The virtual photon as a gauge boson model is, of course, garbage. We shall see that the zepton model of electricity and magnetism is mathematically equivalent to the classical model with its old field theories. Ultimately the entirety of classical electromagnetic theory can be explained from first principles using the zepton model. By finally understanding the underlying physical mechanism rather than a nonsensical virtual photon model, we now have deeper insight into the workings of the zero-point universe.

Here is a list of key points from this chapter.

21) The electric field is zeptons oriented with respect to charge
22) The magnetic field is zeptons rotating in response to charge motion
23) The zero-point field provides both the medium and means for all electromagnetic force interactions
24) Photons are not gauge bosons for electromagnetic forces
25) Zeptons are induced into existence such that they conserve electric and magnetic fields
26) There are no magnetic monopoles

[40] P.A.M. Dirac, 'The Evolution of the Physicist's Picture of Nature", Scientific American, May 1963, 208(5), 45-53.

[41] J.C. Maxwell, "On Physical Lines of Force" Part I "The Theory of Molecular Vortices applied to Magnetic Phenomena" Philosophical Magazine XXI, 1861. pp 161-175. Part II "The theory of Molecular Vortices applied to Electric Currents" Philosophical Magazine XXI, 1861 pp 281-291, 338-348. Part III "The Theory of Molecular Vortices Applied to Statical Electricity" Philosophical Magazine XXIII, 1862, pp 12-24. Part IV "The Theory of Molecular Vortices Applied to the Action of Magnetism and Polarized Light" Philosophical Magazine XXIII, 1862 pp 85-95

[42] J. Larmor, A dynamical theory of the electric and luminiferous ether, Phil. Trans., vol. 185A 1894

Chapter 7: Maxwell's Equations

To illustrate the action of molecular vortices, let s n be the direction of the magnetic force in the field, and let C be the section of an ascending magnetic current perpendicular to the paper. The lines of force due to this current will be circles drawn in the opposite direction from that of the hands of a watch; that is in the direction n w s e. At e the lines of force will be the sum of those of the field and of the current and at w they will be the difference of the two sets of lines; so that the vortices on the east side of the current will be more powerful than those on the west side. Both sets of vortices have their equatorial parts turned toward C, but those on the east side have the greatest effect, so the resultant effect on the current is to urge it towards the west.[43]

James Clerk Maxwell, 1861

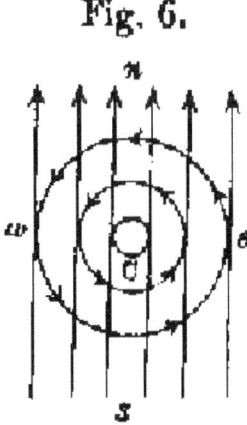

Figure 7-1 Figure 6 From Maxwell's "On Physical Lines of Force" Part I "The Theory of Molecular Vortices applied to Magnetic Phenomena" described in the above quote.

On Physical Lines of Force

The electromagnetic force is described by a system of equations first combined into a coherent theory by James Clerk Maxwell and originally published in his four-part paper "On Physical Lines of Force." He describes each force interaction as being part of a "theory of molecular vortices." The use of a molecular vortex concept to provide a physical transmission mechanism was revived by a contemporary of Maxwell's, William Rankine, who attempted to use them as a means of explaining the conduction of heat. While Maxwell had his own ideas on the conduction of heat he found molecular vortices to be a useful visualization approach for imagining the action of magnetic forces.

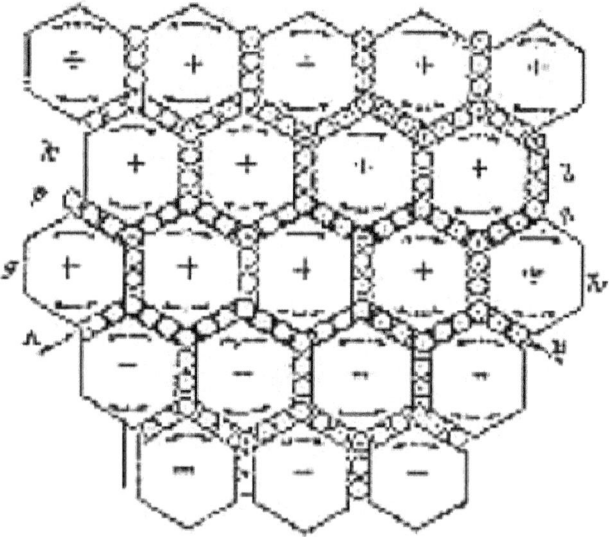

Figure 7-2 Figure 2 From Maxwell's "On Physical Lines of Force" Part II "The Theory of Molecular Vortices applied to Electric Currents" showing a current moving from A to B causing the adjacent vortices to spin in opposite directions. The + and – signs indicate direction of spin rather than charge. After considering the induced currents and rotations of vortices on page 291 he concludes with the below quotation.

There is little doubt that Maxwell's method for physically representing magnetic forces contributed greatly to his ability to combine the theories of electricity and magnetism into one coherent theory. And, while he commented on the dipolar nature of the vortices, he did not expand on that idea to consider them as true particle dipoles. This was after all still 36 years prior to JJ Thompson's discovery of the electron, 50 years prior to Plank's quantum harmonic oscillator, and 71 years before the discovery of the positron. He unfortunately did not see that static electric currents could be simply considered to be non-rotating electric dipoles, as he always referred to them as rotating vortices. This shortcoming is perhaps why the molecular vortex model of electromagnetic theory did not survive to the present day. Only now with zero-point field theory firmly established can we recognize that Maxwell's vortex theory of magnetism was essentially correct.

> *It appears therefore that the phenomena of induced currents are part of the process of communicating the rotary velocity of the vortices from one part of the field to the other.* [44]

James Clerk Maxwell, 1861

To understand the difference this model makes to one's understanding one need only consider the simplest of interactions, such as between a moving charged particle and an electric or magnetic field. In the simplest case of an electric field interaction as shown in Figure 7-3 we have a motionless, positively charged body, which is surrounded by zepton dipoles with their negative charges oriented toward the body and their positive charges oriented away. As another positively charged particle approaches it will also have zepton dipoles oriented around it. As the second particle moves toward the first the two fields of zeptons will interact, repelling each other building up pressure in between. The increased pressure of the vacuum deflects the moving positively charged body away from the static charge. A moving negative charge is deflected in the opposite direction, as the pressure between two like charges is lower than the natural background pressure.

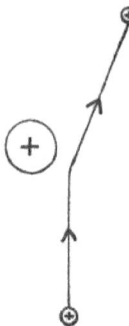

Figure 7-3 A simple static electric interaction between a small moving positively charged body and a larger positively charged body.

The electrostatic example is intuitively quite simple and easy for most people to understand. In contrast, the magnetic force interaction is less intuitive, as it is described as a vector cross product. The magnetic field is in one direction. The motion of the charge is perpendicular to the magnetic field and the direction of the force that makes the charge move is perpendicular to both. This is often taught together with a visual mnemonic aid the right hand rule, invented by John Ambrose Fleming. This rule is different from the right hand grip rule used before.

> *It is important that the student should, even at this stage, realize that the mechanical actions occurring between magnetic poles are not due to action at a distance, as it has been called, or to magnetic poles pulling or pushing other magnetic poles across empty space without intermediary machinery; but they must be regarded as the visible effects of operations taking place in a medium called the electromagnetic medium or ether, which fills all space. The mutual dynamical action of magnetic poles can be accounted for by the assumption that this medium,*

when traversed by magnetic flux, tends to contract or shrink along the direction of the lines of magnetic flux, and tends to expand or swell out in a direction at right angles to them.[45]

John Ambrose Fleming, 1902

To use the right hand rule, first position your hand as shown in Figure 7-4. The second finger points in the direction of the magnetic field, the first finger points in the direction of motion of the particle, and the thumb points in the direction of the force on the particle. The force on the body deflects it in the direction the thumb is pointing. This is not the most intuitive scheme that physicists could come up with, as your first inclination may be to think that the charged particle would be pushed in the direction of the magnetic field instead of in a direction at 90 degrees to the magnetic field.

Figure 7-4 Placement of fingers for the right hand rule memory device.

The physical description of the magnetic field as rotating zepton dipoles gives us a clearer picture. The magnetic field direction now just tells us the direction of spin of the dipoles, not to be confused at all with the direction of force. The rotation of the dipoles is now in the same plane as the body moving through the field as seen in Figure 7-5. A positive charge is deflected to the left by a clockwise spinning dipole field, a downward directed magnetic field. If we imagined that the charge to dipole interactions where strictly mechanical, it is easy to see that the clockwise rotation sweeps the charge to the side.

Figure 7-5 A simple electromagnetic interaction between a small moving positively charged body and a rotating zepton dipole.

Unfortunately it is not that simple as negative charges deflect the opposite direction. Using the right hand rule we have to point the first finger in the opposite direction from motion to get the correct deflection. Indeed it appears that mechanically we have to think of a negative charge acting like a positive charge moving backwards through time, or perhaps as a kind of negative energy, not at all unlike the negative energy associated with antimatter as described by Dirac's equation. In any case it is much easier to visualize what happens in magnetic interactions using Maxwell's vortex model, or the zepton model.

We can more easily understand how electric attraction or repulsion between a moving charge and the zepton dipoles causes the charge to be deflected and be swept to the side. Keep in mind that as with all other forces discussed so far, it is ultimately a pressure differential that produces movement. With the underlying physical mechanism better understood, we can now say that classical electricity and magnetism meets the criteria as an acceptable force model for describing the zero-point universe, that is, at least on scales of distance where classical theory is appropriate.

Maxwell's Equations

As a matter of completeness we can briefly address Maxwell's equations in their modern integral form, as they are the four equations that describe electromagnetic theory in their entirety. The first equation is Gauss's Law. This states what was mentioned earlier, that the sum of the electric field around a charge is not dependent on distance. For any area **A**, such as a sphere, around a body of charge the total energy of the electric field **E** is constant and related to the total charge **Q**. In terms of zepton dipoles it means essentially that the total amount of dipole orientation will be the same taken over any area around the charge. The constant ε_0 is the permittivity of the vacuum, which equals approximately **8.854 x 10^{-12} C^2/Vm** (Coulomb squared per volt-meter). It can be expressed in other units such as Coulomb squared per Newton meter squared (C^2/Nm^2). Gauss's Law is shown in Equation 7-1 in its simple integral form.

Equation 7-1

$$\oint E \cdot dA = \frac{Q}{\epsilon_0}$$

Gauss's Law is essentially a restatement of Coulombs Law, or *vice versa*, so only one of them appears in Maxwell's equations, with the more general form being Gauss's Law. Coulomb's Law is in a form that is often more practical for calculating electrostatic forces **F** since it has both charges, **Q$_1$,Q$_2$**, and the distance **r** in the formula. The **4π** factor relates the equations, as **4πr^2** is the surface area of a sphere around the charge. Coulomb's Law is shown in Equation 7-2.

Equation 7-2

$$F = \frac{Q_1 Q_2}{4\pi \epsilon_0 r^2}$$

The second of Maxwell's equations is Gauss's Law for Magnetism as shown in simple integral form in Equation 7-3. This Law is basically a statement that an area A enclosing a source of magnetism B has a total magnetic field of zero. As we discussed previously the same amount of magnetism emanates from the north pole of a magnet as emanates from the south pole. Gauss's Law for Magnetism is sometimes stated as the law that there are no magnetic monopoles. In a spinning zepton dipole model for magnetism we can easily see that the magnetic field above each dipole is equal but opposite to the magnetic field below it. If you put a bunch of rotating zepton dipoles together, the magnetic field always adds up to zero.

Equation 7-3

$$\oint B \cdot dA = 0$$

The third equation is Faraday's Law of Induction. This law is most commonly stated as; the electromotive force in any closed circuit is equal to the time rate of change of the magnetic flux through the circuit. This law in the simplest case relates to a conductor moving a distance l through the flux of a magnetic field Φ_B. The faster (time = t) the conductor is moved, the faster the rate of change, and the higher the electromotive force that is generated. With the electromotive force being measured in volts. Higher electromotive forces lead to proportionally higher currents through the circuit.

Equation 7-4

$$\oint E \cdot dl = -\frac{d\Phi_B}{dt}$$

The last of the standard four equations is Ampere's law with Maxwell's correction. Ampere's law was originally a line integral $\int dl$ around a magnetic field B being equal to the current I in a conductor multiplied the permeability μ_0 of the vacuum. To put it simply, when the current in a conductor increases, the magnetic field increases too. Note that μ_0 = **4π x 10⁻⁷ Vs/Am** (Volt Second per Amp meter), which can be restated in a variety of other useful units.

Equation 7-5

$$\oint B \cdot dl = \mu_0 I$$

Maxwell recognized some shortcomings in Ampere's Law and came up with what is called displacement current, **dΨ_E/dt** using his molecular vortex model. The displacement current term is related to the polarization of the medium of the vacuum, the electric field or as we now know it to be, zeptons. He added the displacement current, the rate of change of polarization of the vacuum to the current in the conductor to obtain the total magnetic field. This gives us the final equation of the four principle equations in electromagnetic theory as shown in Equation 7-6. Note that all these equations can be expressed in many different forms, including differential forms and various other integral forms.

Equation 7-6

$$\oint B \cdot dl = \mu_0 \left(I + \epsilon_0 \frac{d\Psi_E}{dt} \right)$$

With that we have tackled one of the Standard Model forces and found a way to make sense of it in light of the existence of zero-point energy, while incorporating a mechanical explanation for force transmission and the movement of bodies. We have determined that Maxwell's molecular vortex theory of magnetism was correct at least with respect to the magnetic part of his theory. The vortices are rotating zepton dipoles.

In this chapter the following points have been shown:

27) Maxwell's vortex theory of magnetism is a precursor to the zepton theory
28) The zepton theory of electricity and magnetism is consistent with Maxwell's Equations

[43] J.C. Maxwell, "On Physical Lines of Force" Part I "The Theory of Molecular Vortices applied to Magnetic Phenomena" Philosophical Magazine XXI, 1861. pp 161-175.

[44] J.C. Maxwell, "On Physical Lines of Force" Part III "The Theory of Molecular Vortices Applied to Electric Currents" Philosophical Magazine XXI, 1861 pp 281-291, 338-348.

[45] J. A. Fleming, Magnetic and Electric Currents, 2nd Edition (London: E. & F. N. Spon, Limited; New York: Spon & Chamberlain, 1902). Pg 19

Chapter 8: The Mattermagnetic Field

Descartes' vortex theory of planetary motion proved initially to be one of the most influential aspects of Cartesian physics, at least until roughly the mid-eighteenth century. A vortex, for Descartes, is a large circling band of material particles. In essence, Descartes' vortex theory attempts to explain celestial phenomena, especially the orbits of the planets or the motions of comets, by situating them (usually at rest) in these large circling bands. The entire Cartesian plenum, consequently, is comprised of a network or series of separate, interlocking vortices.[46]

The Stanford Encyclopedia of Philosophy, 2005 & 2009

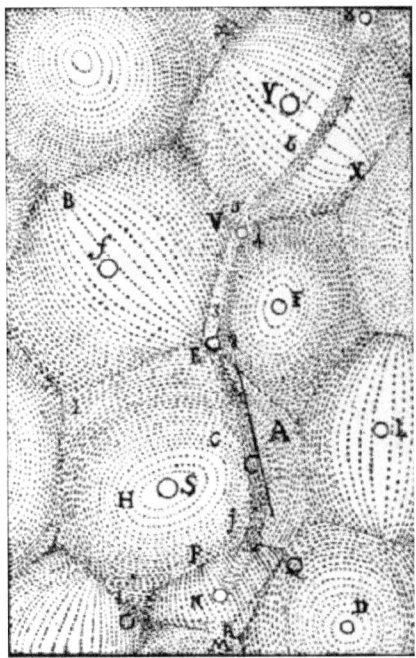

Figure 8-1 An illustration of vortices by Descartes from his *Principia Philosophiae* written in1644.

Descartes's Vortices

Rene Descartes was one of the greatest philosophers in history. He also put forth a great deal of effort trying to understand the universe, in particular the motions of planets. Solely on the basis of intuition and logic he concluded that orbital motion had to be due to a rotation in the vacuum of space, which he termed vortices. Now compared to Maxwell's molecular vortices, Descartes's concept of vortices was different. His were very large such that the planets rode along their edges, oddly enough not that unlike Einstein's view of planets riding along the curvature of space-time. Descartes's theory of large vortices was soundly debunked by Newton in Newton's own *Principia*, and the vortex theory was largely forgotten, except as a mildly interesting historical footnote. After seeing how magnetism works from the perspective of Maxwell, however, Descartes's vortex theory does not seem quite so laughable, at least not when one presumes they are innumerable and small molecular vortices, or as we have now come to know them as rotating zepton dipoles.

Tops and Torques

To begin our exploration of a zepton theory of mechanical interaction we should initially set aside discussion of planets and start with simple rotating objects. The common mechanical energy storage device with the highest energy density is the flywheel, so that should be the first place we look for a zepton-induced force on matter. But, why not start with something more familiar, a top. If we visualize a top when it is not spinning, and stand it on its tip, it will fall over in one direction. There is a point-to-point force, gravity, between the center of mass of the top and the center of mass of the Earth and it will follow that line albeit slightly diverted sideways due to the friction between the base of the top and the surface it is on.

So now, when the same top is spinning it should fall over on its side the same as before and only when it hits the surface should it start rolling around. This is what laws of gravity tell us should happen. Is that what happens? No, the top begins to precess and does not fall over all the way but slowly falls as the top's spinning slows down.

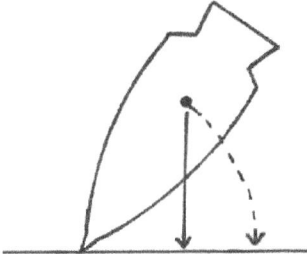

Figure 8-2 A stationary top with the center of mass shown as a dot, the force of gravity a straight solid line arrow, and the falling motion shown by a dashed line arrow

If the top spins one direction, it precesses in that direction and if the top is spun the opposite direction, it processes in the new direction. Where is the spin component of force in Newton's gravity or General Relativity? I must have missed that lecture. What is holding the top up and keeping it from falling? It isn't air; the top does the same thing in a vacuum. It isn't electromagnetic; there are no electromagnetic charges or fields. Also, the center of the gravitational force is no longer at the center of mass as it has shifted to a perpendicular line that goes through the tip of the top where it meets the surface. Shifting of the effective center of weight of an object is not described by Newtonian gravity. The center of mass cannot shift at all. How about Newton's Third Law? Where is the equal but opposite reaction? There is also the thorny problem that a precession is in a circle. It takes a lot of energy to change the axis of rotation of a spinning object, and if the change is a circle then we are not talking about a single change carried forward by momentum. A force must be applied to the top continuously by some means, a very strong force.

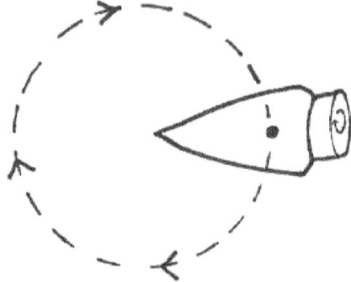

Figure 8-3 A spinning top rotating clockwise as viewed from above with the center of mass shown as a dot, the force of gravity is straight down, but the top precesses around the point where the tip meets the surface instead of falling straight down. The precession is also in a clockwise direction.

Newton's Law of Gravitation is below in Equation 8-1. The force **F** equals Newton's Gravitational constant **G** multiplied by the masses of the two bodies interacting **m_1** and **m_2**, divided by the distance between them, **r** squared.

Equation 8-1

$$F = G\frac{m_1 m_2}{r^2}$$

Newton's Gravitational Constant is experimentally determined and has the value, **G = 6.674 x 10^{-11} m^3/kg s** (meters cubed per kilogram-second). It is similar to Coulombs Law, Equation 7-2, in that the force is dependent on the property of the two bodies that are interacting, divided by the distance squared, multiplied by a constant. The principle difference is that the property in Coulomb's case is electric charge, but in Newton's case it is mass. The other difference is that there is no negative mass, and masses attract while like charges repel. If we think of mass being charge-like that would mean that electric and gravitational forces work in the opposite direction when the charges are the same.

In both cases the force acts linearly between the centers of charge of the two bodies. In Newtonian Gravity there is no field, no flux, and no term that is dependent on rotation. Newtonian gravity cannot explain the precession of a top.

General Relativity does not explain the precession of a top either. General Relativity does include provisions for bodies to follow a curve based on the idea that the presence of mass-energy causes space-time to curve and the body simply follows the curvature. When we discuss a top, however, the energy of rotation is insignificant compared to the energy tied up in the mass of the top, so it does not increase the gravitational force directed toward the earth. Also the curvature of space-time would still be directed toward Earth, not subscribing a circle hovering above it. In the case of the top there should be no measurable difference between the Newtonian computation and the General Relativistic one, and neither account for the precession of a top directly.

Furthermore, the precession of a top is not accounted for by any of the four forces of the Standard Model. It is not electrically charged, so electricity and magnetism do not come into play, and it is not due to the weak force or strong nuclear force. There is certainly a fifth force at work. Oh, that is not to say that it is something entirely new, as there have long been mechanical descriptions of the motion of the top. Those descriptions from Newtonian mechanics are simply not part of any of the four fundamental force theories.

In classical physics, as can be found in any college freshman physics text, the top is described as having angular momentum. The amount of momentum depends on each discrete mass element's distance from the axis of rotation. Alternatively, a top is described as having torque, the moment due to a force at some distance from the axis of rotation. Torque Γ can be generalized as a vector cross product between directional vector r and force F as shown in Equation 8-

2, where **r** is the distance from the rotational axis and **F** is the force perpendicular to **r**. The torque component of the force spinning the top is what we are told keeps the top upright.

Equation 8-2

$$\Gamma = r \times F$$

As before with a vector cross product, we can use the same memory device based on the right hand rule. In this case, the second term of the cross product, the force **F**, points in the direction of the second finger, the first term of the cross product **r** points in the direction of the first finger, and the thumb points in the direction of the resultant torque.

Figure 8-4 Placement of fingers for right hand rule memory device.

Angular momentum is similarly a vector cross product. The angular moment **L** depends on the distance from the center of rotation **r** and the linear momentum **ρ,** where **ρ** is equal to the mass **m** times the velocity **v.** The relationship is shown in Equation 8-3.

Equation 8-3

$$L = r \times \rho = r \times mv$$

Angular momentum can also be expressed as **L** = **I**ω, where **I** is rotational moment of inertia and ω is angular velocity. Note that for a top spinning clockwise the angular momentum vector **L** points downward, while if a top is spinning counter-clockwise the angular momentum vector **L** is pointing up. This is simple to remember with our old friend the right hand grip rule. If you make a fist with your right hand with the fingertips pointed in the direction that the top is rotating, the thumb will point in the direction of the angular momentum.

With that we can finally get to precession. Precession is an effect that is due to the acceleration due to gravity, but not gravity directly. The gravitational force produces another torque **τ** that is proportional to the mass of the top **m**, the acceleration due to Earth's gravity **g**, multiplied by the distance between the tip of the top and a point on the surface directly below the center of mass, which can be determined by taking the distance **r** from the tip to the center of mass and multiplying it by the sine of the angle **a** between the perpendicular and the axis of rotation of the top. Equation 8-4 shows the torque due to gravity.

Equation 8-4

$$\tau = mgr sin\alpha$$

Figure 8-5 illustrates where each component of that equation comes from.

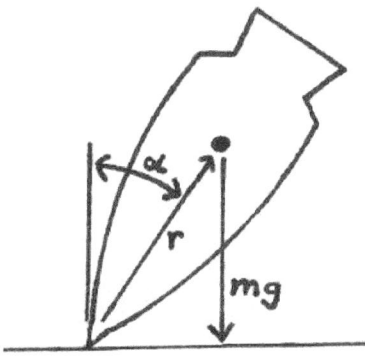

Figure 8-5 A spinning top showing the acceleration due to gravity and the angle between the top's axis of rotation and the perpendicular.

If the force is downward, why does the top take off in a different direction? That is the question. A physicist would say because it is due to another vector cross product, see Equation 8-5, which can be readily solved to give the angular velocity of precession. The toque τ is from Equation 8-5, ω_p is the angular velocity of precession, and L is the angular momentum. Then all you need to do is use the right hand rule to tell you which way the top will precess.

Equation 8-5

$$T = \omega_p \times L$$

Really, is that all we get for an explanation? This is the explanation physicists have accepted for centuries. Certainly it works out mathematically, but it still begs the question, if there is an equal and opposite reaction related to this force, what is the top pushing against in order to stay upright? And, is there a force or not? Since there is nothing there but air and vacuum, if we insist on a mechanical force model we can only conclude that the top is interacting with the vacuum, with some kind of field. We know that the top behaves the same, even when there is no air, so it has to be something in the vacuum. Beyond that, this behavior is truly indicative of a force of some kind. It should not be something that is stuck onto the four standard model forces as an afterthought.

A Different Explanation?!
So far we have described a model of the spinning top based on angular momentum and torque. To better understand the top's interaction with a field in the vacuum we need a third model, as neither of those incorporate a field component. We get a hint about how to proceed from the torque of a current loop shown in Equation 8-6, $\Gamma = m \times B$ with torque Γ being the result of a vector cross product between the vector magnetic moment m and the magnetic field B.[1]

Equation 8-6

$$\Gamma = m \times B$$

A simple electric coil can be described mathematically both in terms of electromagnetic forces and torque produced by those electromagnetic forces. In this case the torque is a due to a Lorentz Force. The well-known Lorentz Force is shown in Equation 8-7, where F is for force, q is for electric charge, v is velocity, and B is the magnetic field - and yes, it is a vector cross product.

Equation 8-7

$$F = q(v \times B)$$

The Lorentz Force equation basically says that as a charge with a given velocity moves perpendicular to a magnetic field there will be a force on it. This does not mean it moves exactly perpendicular, only that the perpendicular component of the motion will produce a force. This is a limited case of Faraday's Law of induction, which is why it is not one of Maxwell's equations. As we shall see the Lorentz Force is a very important, and it is often more useful when discussing bodies of matter, since they normally move as bodies of a fixed size rather than as a current.

Getting back to our top, we can work backwards saying that the torque involved in the motion of a top can be described as being due to a magnetic-like force, a Lorentz Force, on matter. In this case the magnetic field must be a matter-induced magnetic field, and the charge must be something related to a quantity of matter. We can write this hypothetical matter induced Lorentz Force as shown in Equation 8-8, where F_m is the force, m is some yet to be determined quantity of a property of matter, not to be confused with the m in equation 8-6, B_m is the mattermagnetic field, and v is still the velocity. Note that from here on there are many cases where standard symbols in mechanics are used to mean something else in electromagnetic theory, such as in mechanics I means moment of inertia while in electromagnetic theory I means current, so it is important to keep track of what term is being discussed. From here forward a subscript m indicates an electromagnetic term as it hypothetically applies to matter. Existing terms will be used where possible, as we will see; it is unnecessary to make up new terminology.

Equation 8-8

$$F_m = m(v \times B_m)$$

When the top rotates it has a rotational moment of inertia, which we can think of as being something like the magnetic moment of a coil, but being solely due to matter. Instead of velocity v, since it is moving in a circle it needs to be in terms of revolutions per second over a distance equal to $2\pi r$, which is more commonly stated in terms of angular velocity such that $v = r\omega$. In the example, the mattermagnetic field B_m is produced by the spinning top. To derive the mattermagnetic field, we first need to consider the matter equivalent of Ampere's Law as shown in Equation 8-9, where I_m is the matter current and u_{0m} is the matter permeability of the vacuum.

Equation 8-9

$$\oint B_m \cdot dl = \mu_{0m} I_m$$

The problem with Ampere's Law for matter-magnetism is that it is only useful in rare cases. Most of the time we cannot treat matter movement as a current, except for liquid flowing through a pipe. Since an electric current is the amount of charge per second, 1 Amp = 1 Coulomb per second, the units for matter current would be in matter units per second.

Moving matter produces a magnetic field, spinning zepton dipoles, and there are two important principles of interaction. The first is that the mattermagnetic field increases proportionally with increasing amounts of matter. The second is that the mattermagnetic field increases proportionally with increasing velocity of the matter. Combined, those terms equal the momentum of the body of matter, so the mattermagnetic field is also proportional to the momentum.

In Ampere's Law the velocity term is removed, as the velocity of a current is treated as a constant in the part of the equation dealing with the charge momentum. It is not really constant but the speed of propagation of a current in a conductor is very fast, on the order of the speed of light, so variations were not very noticeable to early experimenters. The velocity term was then integrated into the permeability constant for the material, along with the permittivity constant. It is important to remember this when translating between equations that use current and those that use momentum.

Equation 8-10

$$B = \frac{\mu_0}{4\pi} \frac{q}{r^2} v \times \hat{r}$$

To find a more proper form for determining the magnetic field of a top, we need to start with the magnetic field of a moving charge as shown in Equation 8-10. In this equation $\mu_0/4\pi$ is a constant, q is the charge, r is the distance from the charge to the point where the magnetic field is being measured, v is the velocity and r-*hat* is a unit vector in the direction to r. The direction of the magnetic field can then be determined by the right hand rule. This is a form of the Biot-Savart Law for a point charge.

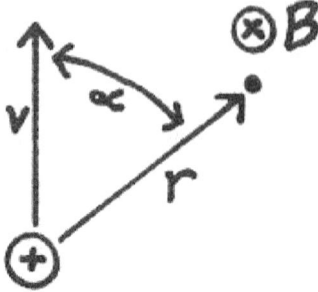

Figure 8-6 A positive charge moving upward produces a magnetic field measured as indicated by the distance vector r. The magnetic field B points into the page on the right side of the moving charge.

Figure 8-6 shows a detail along with the angle that applies to the cross product. Equation 8-11 shows the same equation as 8-9 but with the sine of the angle **a** substituted for the vector cross product.

Equation 8-11

$$B = \frac{\mu_0 q v}{4 \pi r^2} sine \ \alpha$$

In many cases, bodies of matter are too large to be treated as point sources unless the magnetic field is being measured a long way away. In this case we have to look at the sum of the magnetic field produced by each point of matter and integrate over the whole. In the general case the following Equation 8-12 applies. The symbol **u_{0m}** is the matter permeability of the vacuum as before and **d_m** refers to the small point-like pieces of matter being summed.

Equation 8-12

$$B_m = \frac{\mu_{0m}}{4\pi} \int \frac{v \times \hat{r}}{r^2} dm$$

Even simple objects such as our top can lead to relatively complex calculations using the above equation. In cases where it can be treated as a point source the integral can be removed so that we have the matter charge *m* representing the total charge of the body of matter giving us Equation 8-13.

Equation 8-13

$$B_m = \frac{\mu_{0m} m}{4\pi} \frac{v \times \hat{r}}{r^2}$$

We can see clearly now in the numerator *mv*, the momentum of the body of matter, just as we expected. Unfortunately we cannot describe a matter-induced magnetic field around our top with this formula, as we are too close to it for that purpose. In order to simplify things we can think about a rotating cylinder, where all the matter is in the walls of the cylinder and is all some distance *r* from the axis of rotation. If we are looking for the magnetic field along the axis of rotation, we can see that the direction of motion of the cylinder *v* is always perpendicular to the vector pointing toward the axis *r*, so *v x r-hat* = *v*. With any rotating body it is better to express the velocity as an angular velocity such that *v* is replaced by *rω*, where *ω* is the angular velocity an *r* is still the radius of the cylinder. This gives us Equation 8-14 for the mattermagnetic field of a rotating cylinder.

Equation 8-14

$$B_m = \frac{\mu_{0m} m \omega}{4\pi r}$$

We can look at it differently and obtain the same result in a more general way by using the angular momentum and rotational moment of inertia in the equation. Remembering back to Equation 8-3, we can recall that angular momentum *L* is the cross product between the distance *r* and the momentum *mv* and also equal to the product of the angular moment of inertia I multiplied times the angular velocity ω. Both L and I represent the same type of piece-wise integration we are talking about in Equation 8-12, so we can substitute them into the equation as shown in Equation 8-15. Note though that we have to change *r-hat* to *r* putting an additional *r* in the denominator.

Equation 8-15

$$ B_m = \frac{\mu_{0m} L}{4\pi r^3} = \frac{\mu_{0m} I \omega}{4\pi r^3} $$

Then we can substitute in the moment of inertia for a cylinder $I = mr^2$ and obtain Equation 8-14 once again. More generally we can substitute the moment of inertia for any rotating body and obtain a matter-induced magnetic field B_m.

We can now see that by using standard magnetic theory, and applying it to a clockwise spinning top, it can be shown to produce a mattermagnetic field directed downward along its axis of rotation. A top spinning in the counterclockwise direction as viewed from above produces a mattermagnetic field pointing upward. The right hand grip rule can be used to remember the direction of the magnetic field, by curling the fingers in the direction of rotation.

Next, in order to understand precession, we return to the Earth's gravity as it accelerates a slightly off-balance spinning top downward giving it a velocity directed toward Earth. This produces a Lorentz Force, $F_m = m\ (u \times B_m)$. Since the force is a result of a cross product, the movement of the top is perpendicular to the motion due to the gravitational acceleration. The top takes off at a 90-degree angle to the acceleration due to gravity and begins to precess.

This shows that we absolutely can express the motion of a spinning top in terms of a mattermagnetic force interaction. The spinning top produces a mattermagnetic field, and it is the mattermagnetic field that prevents it from falling over. It is the mattermagnetic field that is the thing the top pushes against. It is not pushing against something magical, as the standard explanation would leave you to believe. More than that, as we will soon see, the field is composed of rotating zepton dipoles just as it is with electrically induced magnetic fields.

The mechanics of a top have traditionally been in an odd position as a pseudo-fundamental force along with much of the rest of Newtonian Mechanics. It has been taught as a basic mechanical interaction in any decent mechanics text, but is not included as a component of any of the four fundamental forces of the Standard Model. When one studies electrically induced magnetism of a simple circularly shaped inductor, the concept is presented that the magnetic field can be considered to produce a torque and calculations done accordingly. Well the opposite is also true, a torque due to a rotating mass can also be expressed as result of a magnetic field, and more properly should be. Torque of a top as we know it so far is an expression of matter-magnetism without a linear Coulomb-like force to go with it, yet. Mechanical torque needs to be recognized as a part of a fundamental force, which leads to the key point of this chapter:

29) The physics of a top can be described in terms of a mattermagnetic force

[46] The Stanford Encyclopedia of Philosophy (2005,2009)
http://plato.stanford.edu/entries/descartes-physics/

Chapter 9: Tops, Flywheels, and Gyroscopes

There are several phenomena that we can observe with a gyroscope that suggest that it might conveniently be treated as analogous to an electromagnetic device.[47]

Eric Laithwaite, 1980

Eric Laithwaite

The skeptics among you will no doubt say, show me something more convincing. To answer, I give you Professor Eric Laithwaite, who in his lectures on gyroscopes presented to the Royal Institution in 1973 and 1974 provided some interesting examples of unusual behavior of flywheels and gyroscopes. [47] If he had only discussed how they behave in a manner consistent with electromagnetic theory, then perhaps he would not have been ostracized and branded a crank. But, in the first lecture he claimed that spinning and processing flywheels weigh less then when they are stationary and did demonstrations that he thought supported his contention that they could overcome gravity. He said they violated Newton's laws of physics, and then implied that flywheels could be used as a source of propulsion. Along the way he used a lot of non-standard terms that only made him seen confused. The proceedings from that lecture were never published in a scientific journal. He did, however, publish the same or similar information in his book *Engineer Through the Looking-Glass*, which begins with the statement "The secrets of the universe lie in things that spin."

Perhaps the most interesting experiment was Laithwaite's large gyro on a tower experiment from the 1974 Christmas lecture.[47] In the experiment he used a large flywheel mounted on a heavy-duty stand. In the center was both a rotating and linear bearing so it could move up and down as well as rotate, and a spring supported the weight. There was a pivot point where the flywheel shaft met the bearing assembly, allowing the flywheel to fall under the acceleration due to gravity. Almost amazingly, the spinning flywheel is able to entirely support its own weight, or more precisely vacuum did, without requiring a counterweight. And, the flywheel does look like it was heavy. When he forcibly accelerated the flywheel by pushing the shaft with his hand in the direction of precession, the entire assembly, flywheel, shaft and bearing, moved upward showing that the rise in the flywheel was not entirely countered by a downward force at the center as one might think off-hand. The effective weight of the flywheel and shaft assembly appeared to be reduced by this new acceleration along the direction of precession. As we have seen before, there was a torque, a vector cross product, but this time the resulting force is in an upward direction. Previously we only had a force that kept a top from falling, but now we have one that makes a flywheel appear to be lighter, with only the vacuum available to support the lost weight. The videos of his lectures are readily available on the Internet and are suggested viewing for the interesting demonstrations, while at the same time his descriptions must be carefully filtered.

At this point we need to be clear that flywheels and gyroscopes continue to rotate on their original axis unless a torque is applied to change the axis of rotation. As we all know, a gyroscope on gimbals will maintain its orientation. It is only when a torque is applied that changes the axis of rotation that we see the gyroscope move in a direction perpendicular to the force being applied. Laithwaite demonstrated that in fact, if you lock the flywheel on a stand such that it cannot tilt downward and then try to accelerate it in the normal direction of precession, it simply topples over. It must be free to tilt downward for precession to occur due to the torque due to gravity. It must likewise be free to tilt upward in order to accelerate it in the direction of precession. If you were to put a counterweight opposite the flywheel so that it was accelerated upward when released, the flywheel would precess in the opposite direction from before.

If you take a toy gyroscope, spin it counterclockwise as viewed with the shaft pointed toward you, and accelerate it to your right, you will find that it naturally wants to tilt with the upper edge of the flywheel moving toward you and the lower edge away. You see, a gyroscope or flywheel is not really opposing gravity; it is rotating about its center axis where the shaft goes through. With a small gyroscope in the hand, this axis of rotation is in the center of the gyroscope. With a top, the top is pivoting around the tip where it touches the surface. With a flywheel on a shaft, the flywheel and shaft assembly rotates around the point where the other end of the shaft is attached to the stand. In the case of Laithwaite's large flywheel, the introduction of the spring has the effect of offsetting the effective pivot point location, so instead of the pivot point being on the stand, it has been effectively shifted to the side opposite the flywheel, into the air. The fact that Laithwaite's flywheel moves up is just a consequence of it having to rotate about the effective pivot point. If we think back to our top, it is also attempting to tilt upward, which is how its effective weight shifts to the pivot point, and why it does not fall.

The problem with trying to harness this motion to oppose gravity is that the flywheel will rise until it is rotating parallel to the Earth's surface where the upward force stops. The net affect of rotating a spinning flywheel about an axis is that you eventually end up back where you started with no net propulsive force. That is, unless you stop the flywheel during the recovery step or use friction to keep from going backwards. And, if you try multiple flywheel devices attached around the same central pivot point, their fields cancel. Unfortunately, Laithwaite did not realize the difficulties and continued to try to trick flywheels into doing work for him for much of his remaining life. The important thing for us, however, is none of this could happen if the rotating flywheel was not producing and interacting with a mattermagnetic field that is identical in principle to an electromagnetic field, but due entirely to the movement of matter. The fact that this field can lift a flywheel in opposition to gravity, if only temporarily, is an extraordinarily important piece of evidence that conclusively demonstrates that the motion of a flywheel is due to a field force.

Wallace, Morgan, and Inductive Matter Forces

Predating Laithwaite by a few years is a patent by Wallace describing a "kinemassic" force, a force that was later confirmed by Morgan.[48, 49] Both experimenters studied how the movement of one powered flywheel or gyroscope affected a second passive one. The key element of the experiments is that when two flywheels are placed in close proximity, the rotation of the first flywheel induces a rotation in the second passive flywheel when there is only an air gap and no electromagnetic or mechanical means of energy transfer other than perhaps vibration. Keep in mind that this effect is extremely small relative to electromagnetic induction, as the velocities are slow compared to the speed of light.

Figure 9-1 Morgan's inductive flywheel experimental apparatus

Morgan reported that when the first flywheel rotated at high speed, the second flywheel rotated briskly in the opposite direction. A drawing of his apparatus is shown in Figure 9-1. The discs were a one-sixteenth of an inch apart, and the smaller disc weighing about two pounds (0.9 Kg) was rotated by a motor rated at 26,500 rpm. When the first flywheel was stopped suddenly, the second flywheel stopped and reversed direction. This field effect is identical in principle to the inductance we see with electromagnetic effect between two coils, except that it is due to the movement of matter instead of electric current. The stopping of the flywheel induces a back EMF of the matter motion variety (MMF?), causing the second flywheel to reverse direction. Morgan went on to say that he had developed a "Field Theory of Mechanics," although he did not go into any details or reference a publication. Wallace also noted that the Matter Force has its own permeability and permittivity coefficients, so Wallace also considered these forces to be part of a more general force theory.

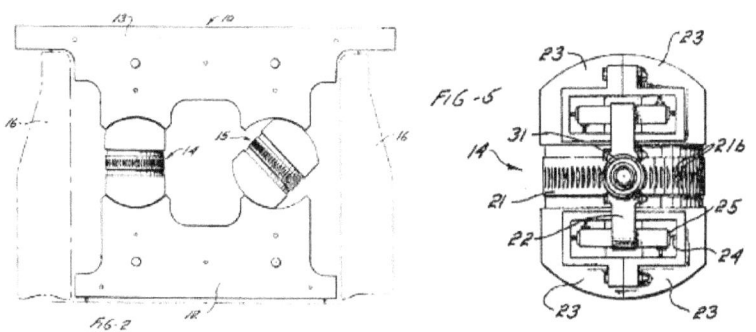

Figure 9-2 Wallace's inductive gyroscope experimental apparatus

One of the interesting differences in their experimental approaches is that Wallace produced the inductive effect by rotating his gyroscopes such that the orientation of the plane of the disc changes with respect to the Earth. This produces a greater force directed in the normal direction of procession, however, his gyroscopes were restrained from moving in that direction. He used a motor and belt drive attached to a pulley, 31 in Figure 9-2, to spin the gyros. The gyroscopes were some distance apart. Morgan simply rotated his flywheels about their axle with the discs oriented perpendicular to the Earth as shown in Figure 9-2. That the inductive force is demonstrated in two different mechanical configurations makes the reports more convincing.

Wallace also claims to produce an upward force opposing gravity with another aspect of his invention. In this case, a ring filled with mercury surrounds the complex multiple rotor gyroscope being rotated as before. It seems highly unlikely that such a force is actually produced. Wallace also spent much of his patent attempting to describe a theory to explain the effect, referring to such things as "spin nuclei" inducing a "secondary time variant gravitational field." Laithwaite's commented, "It is a strange document to say the least."[50] It is, however, the experimental results of inductance that are what is important with respect to this chapter.

Gravitomagnetism

Taken alone, the results of Laithwaite, Wallace, and Morgan are usually dismissed as the work of madmen, but more serious scientific effort has taken place along these lines. Heaviside was the first to propose a gravitomagnetic theory, along the same lines as the kinemassic theory, but with gravity as the operating force back in 1893.[51] This theory was revived and updated as a gravitomagnetic theory consistent with General Relativity by Forward in 1961.[52] Kip Thorne wrote about gravitomagnetics in light of gyroscopic experiments at Stanford.[53] More recently, Tajmar has been a leading advocate for gravitomagnetic theory, even conducting experiments that appear to show fields around rotating superconductors.[54,55]

The work of Laithwaite, Wallace, and Morgan show that there is a magnetic-like force on electrically neutral matter that includes properties of magnetism and inductance. An analysis of the behavior of a precessing flywheel shows that there is an interaction between a magnetic-like force on matter and the vacuum. And that is our key point for this chapter:

30) The principle of magnetic induction applies to mattermagnetic forces

[47] E.R. Laithwaite, Engineer Through the Looking Glass, Ch. 4 The Jabberwock pg 40-57, B.B.C. Publications 1976. Some video clips of his first lecture were part of the BBC show *Heretic*. The BBC broadcast the complete Christmas Lecture, *The Engineer Through the Looking-Glass, Part four:The Jabberwock*.

[48] H. W. Wallace, Method and Apparatus for Creating a Secondary Gravitational Force Field, US Patent No. 3,626,605, issued December 14, 1971

[49] H. Morgan, Now We Can Explore the Universe, IEEE AES Systems Magazine, Vol 13 No. 1 Pg. 5-10 (1998)

[50] E.R. Laithwaite, Roll Isaac, roll – Part II, Electrical Rev. Vol. 204 No. 11 (1979)

[51] O. Heaviside, "A gravitational and electromagnetic analogy" The Electrician 31: 81–82 1893

[52] R.L. Forward, "General Relativity for the Experimentalist," Proceedings of the IRE, 892-586, (1961).

[53] K. S. Thorne, "Gravitomagnetism, Jets in Quasars, and the Stanford Gyroscope experiments" from Near Zero, New Frontiers in Physics, ed. J.D. Fairbank et. al., W.H. Freeman & Co, New York, 1988.

[54] M. Tajmar et al, Coupling of Electromagnetism and Gravitation in the Weak Field Approximation, Physica C 385:551-554 (2003)

[55] M. Tajmar et. al., Measurement of gravitomagnetic and acceleration fields around rotating superconductors, 2006 arxiv.org/pdf/gr-qc/0610015

Chapter 10: Inertia

The vis insita, or innate force of matter, is a power of resisting by which every body, as much as in it lies, endeavours to preserve its present state, whether it be of rest or of moving uniformly forward in a straight line.[56]

Sir Isaac Newton, 1687

The History of Inertia

Inertia is perhaps the single most important principle of any theory of the mechanics of motion. There is good reason why it is Newton's First Law, as it rightly deserves that standing as the first statement that must be made in any physical theory or collection of physical theories. Unfortunately, inertia has never been explained. Even worse, many physicists do not even try to explain it. No universal force theory can ever be considered complete without a fundamental explanation for inertia.

Historically, the first statement of motion that addressed inertia is attributed to Aristotle. However, his statement was a denial of an indefinite inertia, as he thought that a force must be constantly applied to a body in order for it to stay in motion. The mistake can be forgiven since in a world where bodies have to continually overcome friction and air resistance, his statement matches observation. Galileo is credited with the first true statement of inertia as he recognized that "a body moving on a level surface will continue in the same direction at constant speed unless disturbed." In his statement he includes friction and air resistance as disturbances. Newton's First Law is essentially a restatement of Galileo's principle of inertia.

For whether there be any intrinsically material inertia or not, there certainly is an electrical inertia. [...] Quite possibly there is no other kind. Quite possibly that which we observe as the inertia of ordinary matter is simply the electric inertia, or self-induction, of an immense number of ionic charges, or electric atoms, or electrons. This is by far the most interesting hypothesis, because it enables us to progress, and is definite. The admixture of properties – partly explained, viz. the electrical, partly unexplained, viz. the material – lands us nowhere.[57]

Oliver Lodge, 1906

A Physical Force Model

Given what we have covered so far, it should be obvious that inertia can only be an interaction between matter and the zero-point field. We can revisit an image, Figure 10-1, of rotating zeptons interacting with a current in a wire. In the electromagnetic case, the current causes the zeptons to rotate, and the zepton rotation causes the current. The two are inseparable. This is how inertia works with respect to a current loop. As long as there is an electromotive force to overcome resistance in the circuit, the current will continue. But as with Aristotle, it is improper to say that force is required in the absence of disturbances such as resistance. In a hypothetical resistance-free circuit, the current would continue indefinitely. And if there were no current initially, no current would occur without an electromotive force being applied.

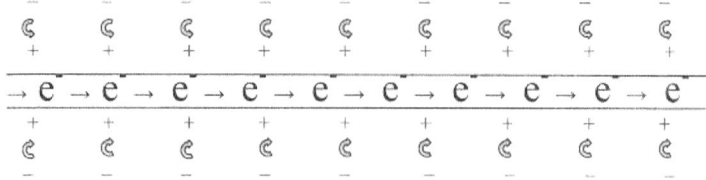

Figure 10-1 Electrons in motion along a line induce rotation in surrounding zeptons.

It has been proposed by Haisch, Rueda, and Puthoff that inertia is due to a Lorentz Force interactions, of the type $F = q (E + u \times B)$ with the zero-point field, so this approach is not entirely new.[58] They develop their theory starting with the electric and magnetic fields generated by the zero-point field, but without relying on strict interpretation of a dipole model for zero-point energy. The real problem is that they only have electric and magnetic forces to begin with and try to relate these forces to an interaction with mass. The effort is not very convincing, as we are left with the question of what the force is between electrically neutral matter and the electric and magnetic fields from the vacuum fluctuations.

> *Lord Kelvin has been the promoter and developer of a view by which the elastic forces between parts of such a medium may be to some extent got rid of as ultimate elements, and be explained by the inertia of a spinning motion of a dynamically permanent kind, which is distributed throughout its volume. If we imagine very minute rapidly-spinning fly-wheels or gyrostats spread through the medium, they will retain their motion for ever, in the absence of friction on their axles, and they will thus form a concrete dynamical illustration of a type of elasticity which arises solely from inertia; and this illustration will be of great use in realising some of the peculiarities of a related type, which I believe can be thoroughly established as the actual type of elasticity transmitting all radiations, whether luminous and thermal or electrical—for they are all one and the same—through the ultimate medium of fluid character of which the vortices constitute matter.*[59]
>
> Joseph Larmor 1894

The real answer is simpler. As a body of matter moves through space it induces zeptons to rotate as shown in Figure 10-2. The number and/or energy of zeptons that rotate is proportional to the momentum, matter-charge times velocity *mv*, of the body of matter. Rotating zeptons also induce matter to move. Inertia is nothing more than the principle of magnetism and inductance applied to the movement of matter. So once again, there must be a magnetic-like field produced by the movement of matter: This time it is required in order to explain inertia.

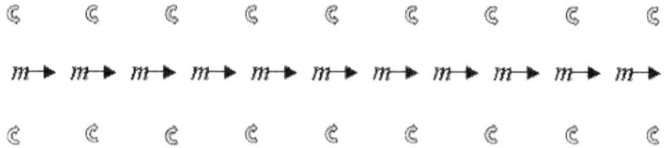

Figure 10-2 Matter in motion along a line inducing rotation in surrounding zeptons.

If the motion of matter can cause zeptons to rotate, they also must cause zeptons to orient themselves with respect to matter when matter is at rest, as shown in Figure 10-3.

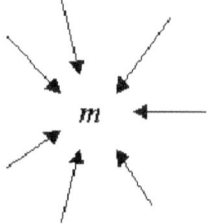

Figure 10-3 A body of matter at rest with a matter field pointing toward it.

It is the matter field or oriented zeptons in a body's rest frame that cause the body to stay at rest. A force must be exerted on the body of matter in order to make it move, and the resistance to motion is provided by the zeptons, as they are forced to rotate when the body moves.

We can briefly note that it is the zeptons that also limit the speed of matter. The energy and time limitations of Heisenberg's uncertainty principle demand that a zepton can only rotate 180 degrees at its maximum extent of travel, its wavelength. So it can only cover a distance equivalent to the speed of light. Since zeptons cannot go any faster, as a body accelerates to near light speed, increasing momentum means an increasing number of zeptons rotating in response to the motion. Perhaps it should more correctly be called the speed of zeptons, since they are the limiting factor.

Many science fiction writers, and science fiction writers posing as physicists, often theorize about faster-than-light travel. Some like to think that a body of matter can be isolated from its inertial field. This concept is more difficult to pull off than they imagine, since matter is porous to the higher energy zeptons, such that bodies of matter are entirely saturated with them. The greatest percentage of a body's inertia is not due to rotating zeptons found off to the side as illustrated in Figure 10-2, but are within the body of matter itself. Therefore, a body of matter cannot be isolated from its inertial field, and cannot go faster than the zeptons can rotate. The speed-of-light issue will be covered in greater detail in coming chapters.

Equation 10-1

$$B_m = \frac{\mu_{0m} m \, v \times \hat{r}}{4\pi} \frac{}{r^2}$$

In order for inertia to exist as an interaction between a moving body of matter and the zero-point vacuum fluctuations, there must be a magnetic-like field generated by moving bodies. This is the same field discussed previously when we were considering a rotating top. The equation for the magnetic field from a moving body is identical in form to that of a moving charge and is the same as shown previously in the last chapter, but is shown here again as Equation 10-1. We can consider this a form of the Biot-Savart Law for matter charge.

We have also introduced the idea that in order for a body to remain at rest, there must be a form of static attraction causing the zepton dipoles to align themselves with respect to matter. This force must take the form of Coulomb's Law or more generally Gauss's Law as shown in Equation 10-2, where M is the matter field and ε_{0m} is the dielectric constant of the vacuum with respect to matter.

Equation 10-2

$$\oint M \cdot dA = \frac{m}{\varepsilon_{0m}}$$

In light of a matter-force theory with a magnetic field produced by the motion of matter and static matter fields produced due to orientation with respect to matter, the Lorentz Force model for inertia becomes an obvious extension of theory. When combined with our earlier examination of the physics of a spinning top, it is becoming increasing obvious that such a force must exist. Newtonian Mechanics, while useful, fails to provide a physical explanation for Newton's First Law, inertia. This is in addition to Newton's mechanics failing to explain the forces behind his Third Law, which requires an equal but opposite force. We are left with the question that since it cannot be electromagnetic, what is the nature of this force? It must be mattermagnetic.

In this chapter it has been shown that:

31) Inertia is a Lorentz Force as it applies to matter

[56] I. Newton, Philosophiæ Naturalis Principia Mathematica, London, 1687

[57] O.Lodge, Electrons or The Nature and Properties of Negative Electricity, George Bell and Sons, London, 1906.

[58] B Haisch, A Rueda, H E Puthoff, Inertia as a zero-point field Lorentz force, Phys Rev A 49: 678 1994.

[59] J. Larmor, A dynamical theory of the electric and luminiferous ether, Phil. Trans., vol. 185A 1894

Chapter 11: The Matter Dipole

Now, if the magnetic state of the fields depends on motions of the medium, a certain force must be exerted in order to increase or diminish these motions, and when the motions are excited they continue, so that the effect on the connexion between the current and electromagnetic field surrounding it, is to endow the current with a kind of momentum, just as the connexion between the driving-point of a machine and fly-wheel endows the driving-point with an additional momentum, which may be called the momentum of the fly-wheel reduced to the driving-point. The unbalanced force acting on the driving-point increases the momentum, and is measured by the rate of its increase.[60]

James Clerk Maxwell, 1864

The Mass and Gravity Models

If this force due to matter is to be fundamentally identical in operation to the electromagnetic force, as it so far appears to be, there must be some kind of charge dipole. Without a charge dipole there cannot be a mattermagnetic force analogous to the electromagnetic force. Wallace and Morgan had independently thought they had discovered a kinnemassic force, a magnetic-like force due to mass, where mass is the charge.[61, 62] The problem with that idea is that there is no negative mass, and without negative mass there can be no mass dipole. Secondly, like charges must repel, and mass is known to be attracted to other masses due to gravity. Oliver Heaviside, followed by others, similarly attempted to develop a gravitomagnetic theory where gravity is the force and the charge is mass or mass-energy.[63,64,65] That theory has similar problems. There is no negative gravity to yield a repulsive force. Like bodies of matter are attracted due to gravity, not repelled. Gravitomagnetic theory also fails on the basis that there is no negative mass. Neither a kinemassic nor gravitomagnetic theory is viable, as there is no charge dipole with either and no repulsive force.

Finding The Real Dipole

In order to figure out the nature of the charge dipole, we need to start by looking at the particle pairs that form the dipoles and search for known properties of the particles that are opposite and could possibly be the previously unrecognized charge dipole. Since virtual electron-positron pairs are the most fundamental dipole, we will start with them. Below is Table 11-1, which lists the basic properties of the two particles in question.

Property	Electron	Positron
Group	Lepton	Lepton
Charge	$-e$	$+e$
Spin	½	½
Mass	0.511 MeV/c²	0.511 MeV/c²
Magnetic Moment	-1 μ_B	+1 μ_B
Matter - Antimatter	Matter	Antimatter

Table **11-1** Comparison of known fundamental properties of an electron and positron

The first property that differs above is electric charge, which we have already dismissed as a possibility, as we need a charge force that works on electrically neutral bodies of matter. While both particles are spin ½, they can be arranged such that they are spinning in opposite directions. In a virtual electron-positron pair, they would have opposite spin. Spin, however, does not yield a charge dipole of the type we are looking for, as large bodies of matter are spin neutral. The next property of interest is magnetic moment as those values are opposite because the electrical charges are opposite. But, magnetic moments are electric in nature and do not interact with electrically neutral bodies of matter in the necessary manner.

116

That leaves us with only one possibility: the opposite charge to matter is antimatter. There simply is no other known property that is dipolar in character. Given only one choice, we are stuck with it. It makes a lot of sense once you think about it for a while. It explains a lot. Like, why is there no antimatter in local space? It is a lot to digest. I personally struggled with the concept for weeks when I first realized that this must be the case, so I know you will too. Nonetheless, I hope you read on.

How could we have missed it? Ah, that is the real question. To start with, matter charge absolutely must be proportional in some way to mass in large bodies of neutral matter. That does not mean equal, just proportional. There is no negative mass, so it must be separate. If electrons and protons each have a matter-charge of one, matter-charge will be proportional to mass, particularly if a neutron has a matter-charge of two. On a large scale, we have already built the mattermagnetic force models into Newtonian mechanics and General Relativity without recognizing it as such. The math does not change very much, only the underlying cause.

Compared to currents and free electric charges that move at or near the speed of light, matter generally moves many orders of magnitude slower. So it has less momentum and, therefore, less energy and less force associated with it, so it gets lost in the statistical noise a lot of the time. Are the dielectric constants of insulating materials strictly due to an electromagnetic interaction, or is there perhaps some small mattermagnetic charge effect being fudged into an electrical one? Oh, there are plenty of ways to hide the force if you think about it, more than any one person could easily catalog. As with Newtonian Mechanics, there has been a systematic approach to incorporate the mattermagnetic forces into other force theories, or even cover up the deficiencies with fudged constants or incorrect theories.

Matter Repulsion

One of the great moments in the history of physics in the last century was when it became commonly accepted that there was Dark Energy, a previously unaccounted for force behind the accelerated expansion of the universe. This was unfortunately immediately followed by one of the low points, wide-scale denial. No, there isn't a missing force we are told; it is manifestation of something else. Many scientists initially even called it the Great Attractor, something nebulous in the far reaches of the universe. This was as if to say, it could not possibly be anything we could detect nearer to us. It certainly cannot affect our gravitational force laws, we are told. What should have occurred was an opening of all Standard Model theories as we conducted a search for anything consistent with a force that causes matter to move away from matter.

The first thing to acknowledge is that matter is moving away from matter. At first glance this should lead us to consider a repulsive force between bodies of matter as the simplest form of force that could account for it, not minding for the moment that gravity is stronger over other distance ranges. To a true classicist, this is not the first indication of an unidentified repulsive force between matter. After all, what is the repulsive force that counteracts the Coulomb attraction between a proton and an electron? Oh, we know the manner in which that question was swept aside, but classically we will not truly understand the hydrogen atom until the mean distance can be described with a force balance equation. An electron is accelerated toward a proton, and there has to be some kind of force that stops it or hydrogen or any other atom would not exist. The bottom line is that we also have an unidentified short-range force causing matter to be repelled from matter. Consequently, it should not be unexpected that there are other instances where matter is repelled by matter.

That gives us both a very long-range and a very short-range force repelling matter from matter. In the spirit of Occam's Razor, we should then consider that these two

118

force manifestations are due to the same fundamental but previously unknown force. Since this proposed force covers all scales of size, we must look for its manifestations on scales of size more common to our everyday experience. We must also acknowledge that this force must have been well disguised to remain hidden beneath electromagnetism and gravity. We have now found this force, the so-called Dark Energy. It is the repulsive Matter Force.

Matter Field Lines

As we have for electricity and magnetism, we can have Faraday Field lines for matter forces and mattermagnetism. And, just as with electricity and magnetism, those force lines are more accurately represented as zepton dipoles. Figure 11-1 illustrates the basic matter field around a body of matter with the antimatter side of the dipoles oriented toward the matter. The zeptons become polarized, and the lines of polarization extend toward infinity.

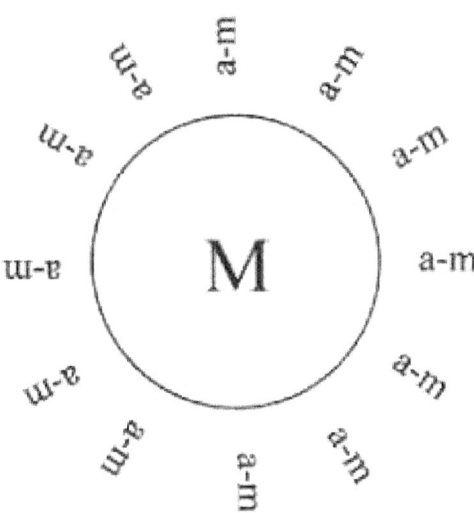

Figure 11-1 A body of matter surrounded by matter-oriented zeptons.

Then if we have two bodies of matter, their lines of polarization will meet, matter to matter, and be repelled as shown in Figure 11-2.

Figure 11-2 Two bodies of matter are surrounded by matter-oriented zeptons in nearby space, oriented in a repulsive configuration.

As with the electrical case, zeptons fill the void, which leads to an increase in zepton pressure in the space between the bodies, pushing them apart. As before with electricity and magnetism, motion is caused by a force differential, as there is also a constant pressure originating from the vacuum pushing the bodies together. The matter attractive field lines will also be analogous to the electrostatic attractive field lines.

Here we need to remember that there is more than one type of zepton dipole. There are those that are electron-like with negative charge and matter on one end and positive charge and antimatter on the other end. If the vacuum were only composed of one type of charge in the above scenario, there would also have to be an electric field to go along with the matter field. Luckily, there are particle pairs with the opposite charge arrangement, with positive charge and matter on one end of the dipole and negative charge and antimatter on the other end of the dipole. These are proton-like zeptons. If an electron-like and a proton-like zepton are side by side, they cancel with respect to one type of charge and are

120

polarized with respect to the other. Figure 11-3 shows a zepton pair polarized with respect to matter charge. They can be similarly polarized with respect to electric charge while being matter neutral. However, in local space dominated by matter, there will always be some matter charge polarization.

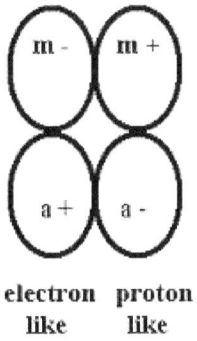

electron proton
like like

Figure 11-3 A zepton pair polarized with respect to matter charge while neutral with respect to electric charge.

One could also speculate that the above configuration is equivalent to a virtual neutron-antineutron particle pair. Such a particle pair could carry matter charge without electrical charge. However, because neutrons are formed when an electron combines with a proton and quickly decays back to a proton and electron it is not entirely clear if a neutron is fundamental enough to form a zepton. This is a question that will remain open pending a better understanding of the particles.

Now, one of the big questions must be stuck in your mind: How can it be that the repulsive force between matter is responsible for the accelerating expansion of the universe while at the same time be weaker than gravity at smaller distances? To answer requires that we complete a first principle derivation of gravity. Suffice it to say for now that once we understand how both gravity and the matter repulsive force propagate through space, it becomes easy to understand how gravity can be stronger over shorter distances while matter repulsion is stronger over longer distances.

Matter Magnetism
As just discussed with respect to inertia, the motion of a body of matter causes nearby zeptons to spin as illustrated in Figure 11-4.

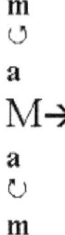

Figure 11-4 A body of matter moving through space causes nearby zeptons to spin.

As before with electricity and magnetism, we can use the right hand grip rule to determine the direction of the mattermagnetic field $\boldsymbol{B_m}$. For spinning zeptons shown in the illustration above, the moving body of matter the magnetic field points out of the page, while the magnetic field below the moving body of matter points into the page.

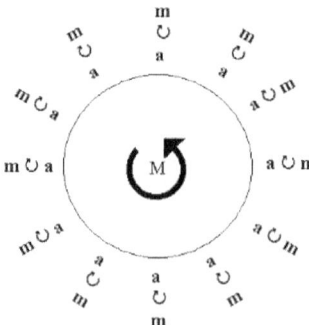

Figure 11-5 A body of matter rotating in space causes nearby zeptons outside its circumference to rotate in the opposite direction.

One aspect of the Matter Force model that achieves greater importance when discussing large bodies of matter, is the force generated when bodies are rotating. A large rotating body produces a counter-rotating mattermagnetic field around its circumference, as illustrated in Figure 11-5.

As before, when electron-like zeptons and proton-like zeptons form pairs, it is possible for them to produce a mattermagnetic field when rotating while not producing an electromagnetic field and *vice versa*. Such a rotating pair is illustrated in Figure 11-6. Note that zeptons are effectively infinitely small compared to the scale of bodies in orbit, so that there is no measurable electromagnetic field in this case. At the Plank scale, 10^{-35} meters, there may be an effect, but that is beyond our observational capabilities. Alternatively, there may possibly be virtual neutron-antineutron pairs, which would generate no electromagnetic field.

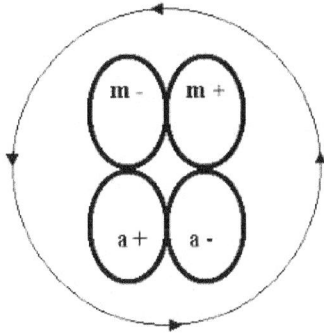

Figure 11-6 A zepton pair polarized with respect to matter charge while neutral with respect to electric charge, so that when they rotate they can produce a mattermagnetic field without producing an electromagnetic field.

Newtonian gravity does not include a rotational force between large bodies of matter, but the Matter Force does. We will see shortly that this deficiency has led to numerous problems being left unsolved, or incorrectly solved with the theory of General Relativity.

Not only did we determine the nature of our charge dipoles and explain force interactions in terms of point-to-point zepton pressure induced forces, we solved the puzzle of the nature of dark energy. In this chapter we have determined the following:

32) The Matter Force charge dipole is matter and antimatter
33) Zeptons provide the means and medium of interaction for the Matter Force
34) Dark Energy is the repulsive component of the Matter Force

[60] J.C. Maxwell, "A dynamical theory of the electromagnetic field" Philosophical Transactions of the Royal Society of London 155: 459–512. doi:10.1098/rstl.1865.0008 1865.

[61] H. W. Wallace, Method and Apparatus for Creating a Secondary Gravitational Force Field, US Patent No. 3,626,605, issued December 14, 1971.

[62] H. Morgan, Now We Can Explore the Universe, IEEE AES Systems Magazine, Vol 13 No. 1 Pg. 5-10, 1998.

[63] O. Heaviside, "A gravitational and electromagnetic analogy" The Electrician 31: 81–82 1893

[64] R.L. Forward, "General Relativity for the Experimentalist," Proceedings of the IRE, 892-586, 1961.

[65] M. Tajmar et al, Coupling of Electromagnetism and Gravitation in the Weak Field Approximation, Physica C 385:551-554 2003.

Chapter 12: Astronomical Motion

Each failure to explain the spiral arms makes it more and more difficult to resist a suspicion that the spiral nebulae are the seat of types of forces entirely unknown to us, forces which may possibly express novel and unsuspected metric properties of space.[66]

James Hopwood Jeans, 1928

Perhaps the Matter Force Theory explaining spinning tops and flywheels, inertia, and Dark Energy is insufficient for some readers. A careful examination of the motion of astronomical bodies illuminates even more convincing arguments that favor the Matter Force Theory. To start with, we will look at a complication, which affected all of physics, or should have.

Attractive and Repulsive Forces

Historically, there has been but a single force to explain the motions of astronomical bodies, the force of gravity. In the Newtonian view, this force is only attractive, somehow pulling bodies of matter toward each other. The interaction is point-to-point, center-of-mass to center-of-mass. To gravity is added Newtonian Mechanics, a quasi-force, which is sometimes used to explain odd behavior that cannot be explained by a strictly linearly acting force. The force behind the expansion of the universe was unknown, as the expansion had long been thought of as some residual velocity due to an explosive event long ago, the so-called Big Bang. The other three fundamental forces of the standard model - electricity and magnetism, the weak force and the strong force - play no role in astronomical motion of large, electrically neutral bodies.

It was 1998 when the complication raised its head, when there was conclusive evidence published that the expansion of the universe was accelerating.[67] It was at that moment that it became a certainty that there was a completely unknown new force, which had to be responsible for the acceleration. That necessity is not disputed. What has been continually overlooked is at that moment in time we had to recognize that Newtonian Gravity and General Relativity had to be a superposition of two forces, one attractive and one repulsive. That conclusion has nothing to do with Matter Force Theory, as any theory that accounts for the acceleration requires it. In the meantime, mainstream physicists that freely admit that accelerated expansion is correct will stick their fingers in their ears like small children saying, "la la la la, I can't hear you" if you bring up the fact that the present theories of gravity are no longer acceptable.

Beyond that, we must also recognize that there is a mechanism responsible for gravity being stronger than the force behind expansion over distances on the scale of galaxies and smaller, while the repulsive force is stronger on larger distance scales. This fundamental complication once again has nothing to do with Matter Force Theory, but is a fundamental problem that must be overcome by any proposed force theory that accounts for the accelerating expansion of the universe. That problem is addressed in a later chapter.

Once one recognizes that Newtonian Gravity must be the superposition of two forces, it is much easier to tackle the problem head-on. First there must be a gravitational force, some kind of super-gravity that pushes bodies together and is stronger than Newtonian Gravity. Then there must be a repulsive force behind the expansion, which is also stronger than Newtonian Gravity, but in the opposite direction. Note that with respect to linear forces, General Relativity is taken here to be equivalent to Newtonian Gravity. The concept is simply illustrated in Figure 12-1, where Newtonian gravity is the sum of the other two real forces. From here on the word "gravity" will refer to the stronger real force, and "Newtonian Gravity" will refer to the weaker force summation, the difference between the two principle forces.

126

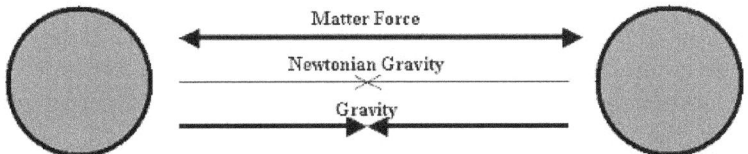

Figure 12-1 Two bodies with the Matter Force pushing them apart, gravity pushing them together, with the net difference being Newtonian Gravity.

The equations are still quite simple. We can start with Newtonian Gravity as shown in Equation 12-1, where the gravitational force F is equal to Newton's Gravitational constant G_N multiplied by the masses of the two bodies M_1 and M_2 divided by the distance r squared taken between the center of mass of the two bodies. This formula is still the simplest to use in general practice in cases where it is applicable.

Equation 12-1

$$ F = G_N \frac{M_1 M_2}{r^2} $$

The matter repulsion force can be expressed in a form comparable to Coulomb's Law as shown in Equation 12-2. In this case the force due to matter F_m is equal to the inverse of $4\pi\varepsilon_{0m}$ where ε_{0m} is the dielectric constant of the vacuum with respect to the Matter Force, multiplied by the matter charges of the two bodies m_1 and m_2 divided by the distance r squared taken between the center of mass of the two bodies.

Equation 12-2

$$ F_m = \frac{1}{4\pi\varepsilon_{0m}} \frac{m_1 m_2}{r^2} $$

Since we do not yet know the source of gravity within the scope of the zero-point universe theory, it is shown in a form similar to Newtonian Gravity, Equation 12-3.

In this case, the force due to matter F_m is equal to the inverse of a gravitational constant G_m multiplied by the matter charge of the two bodies m_1 and m_2 divided by the distance r squared taken between the center of mass of the two bodies.

Equation 12-3

$$F_m = G_m \frac{m_1 m_2}{r^2}$$

As stated in a previous chapter, the matter charge must be proportional to mass of large neutral bodies, while possibly allowing for the constants to differ. Otherwise gravity and the matter repulsive force could not be summed to equal Newtonian Gravity. On the other hand, matter charge is almost certainly not mass, as in order to be completely consistent with electromagnetic forces, an electron and proton need to have the same unit matter charge. In a coming chapter, the mass of the proton and electron are derived from first principles. It is evident from that derivation that mass is not a fundamental property and therefore cannot be a fundamental charge. For the time being we will leave the question of the nature of matter charge open.

Circular Orbits

An approximately circular orbit occurs when a small body orbits a much larger body. The mechanics of such orbits follow, for the most part, the well-known laws of motion established by Newton. In our solar system, these laws work very well, as Newtonian Gravity allows for accurate mathematical computations. There is, however, a smaller effect caused by the rotation of the bodies and transmitted through space by the mattermagnetic force.

When a large body such as our Sun rotates, it produces a mattermagnetic field that points in the opposite direction from the Sun's angular momentum. If the Sun is seen from above as rotating counterclockwise, the zeptons around it will rotate clockwise. The angular momentum of the sun, which is directed toward us, causes a mattermagnetic field B_m directed away from us. As the smaller body crosses this mattermagnetic field, it is subject to a Lorentz Force F_m as shown in Equation 12-4.

Equation 12-4

$$F_m = m_2(v \times B_m)$$

This is better illustrated in Figure 12-2, where a large body is rotating with an angular velocity ω, producing a mattermagnetic field directed into the page. The smaller orbiting body moves with velocity v, which produces a force F_m directed toward the larger body. Note that as long as the larger body is rotating in the same direction as the orbiting body, the resultant force will be directed inward in accordance with the right hand rule.

Figure 12-2 One small body orbiting a larger body. The larger body is rotating, producing a mattermagnetic field of rotation in the opposite direction. As the small body moves through the field in the direction of the velocity arrow, the mattermagnetic force pushes it towards the larger body.

If you think that this extra force makes the orbiting body appear heavier than it actually is, you would be correct. The interesting thing is that we do not have a balance out in space to measure the mass of the planets. The mass has always been inferred based on

Newton's law of gravitation and, more recently, General Relativity. If there is a small error in the mass measurement, that is OK, the mass is still being measured indirectly. The previous results cannot be used to prove anything, as long as the density is reasonable for the composition.

Synchronous Rotation

The next force interaction we should look at is synchronous rotation. Synchronous rotation is best known from studying our Moon. The Moon takes as long to rotate on its axis as it does to complete one orbit, so that the same half of the Moon is always facing the Earth, and the other half is always facing away. This effect is found between other bodies and is often called "tidal locking," because it is said to be due to "tidal forces."

The problem here is that physicists are saying that tides affect the rotation of solid bodies, which makes no sense. Yet when you mention tidal forces, others nod their head like they know what you are talking about. Newtonian Gravity does not have a rotational component; it is a point-to-point linear force. Certainly some odd shaped bodies will wobble, but to actually slow down the rotation of a moon or planet takes a different type of force, one that has a magnetic-like force component. It takes mattermagnetism.

The funny thing is that once we are aware of the Matter Force, explaining the slowing of rotation of orbiting bodies becomes trivial. We can consider a smaller satellite body rotating in the same direction as the larger body it is orbiting in the same direction as its orbit. We will assume that is a counterclockwise direction once again as in Figure 12-2.

A planet rotating counterclockwise produces a magnetic field pointing into the page as shown in Figure 12-3. This is the body's angular inertia. Note that the mattermagnetic field due to a rotating sun as shown in Figure 12-2 also points into the page.

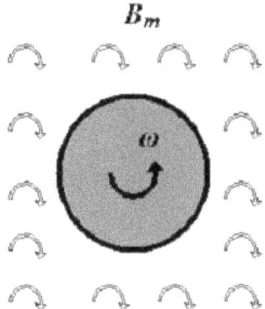

Figure 12-3 A small body rotating counterclockwise produces a counter-rotating zepton field.

Next we need to look at the magnetic field produced by the orbiting body due to its orbital velocity, as shown in Figure 12-4. In this case, the field on the side away from the Sun also points into the page, while the field between the orbiting body and the Sun points out of the page in opposition to all the other fields. The field on the inside opposes the field due to rotation shown in Figure 12-3, while the field on the outside enforces it. In free space, these two effects are equal, so the angular velocity does not change.

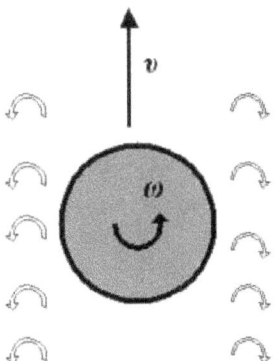

Figure 12-4 A small body moving in an orbit produces oppositely rotating zepton fields on either side of it. The field due to its rotation is not shown.

It is important to note that the small orbiting body's orbital momentum is much greater than its angular momentum due to rotation. The planet's orbital magnetic field is opposed by the Sun's rotational magnetic field in the region of space between them, while on the outer side of the planet, the two fields point in the same direction. Because the field strength of the Sun's rotational magnetic field varies in inverse proportion to the square of the distance, the magnetic force working against the planet's rotation is slightly greater than the magnetic force working with the planet's rotation. This differential causes the planet's rotation to slow.

Keep in mind that the total energy of the system does not change, so as a body in orbit loses angular momentum, it gains linear momentum. Linear momentum means an increase in velocity, and the increased velocity requires that the distance between the two bodies must increase. Note that the Earth's rotation is known to be slowing while the moon's orbital distance is increasing. The action of the force between the two bodies is symmetric. Both the larger and smaller body lose angular velocity over time.

It is baffling that scientists today still accept that changes in Newtonian Gravity or General Relativity due to tides could cause synchronous rotation. The so-called tidal forces between solid rotating bodies have nothing to do with tides, but are a manifestation of the mattermagnetic force. True, there is some tidal action in a body like Earth with a great deal of liquid on its surface, but even in the general case the mattermagnetic field is what is significant. The explanation of tidal interactions is another significant addition to the understanding of the physics of the zero-point universe.

Precession of Elliptical Orbits
The next important experiment already conducted for us is the precession of elliptical orbits. Scientists recognized early on that it was impossible for Newtonian

Gravity to cause precession of an elliptical orbit in a two-body system where both bodies are homogenous spheres. Precession of an ellipse is shown in Figure 12-5, with the precession advancing in the counterclockwise direction.

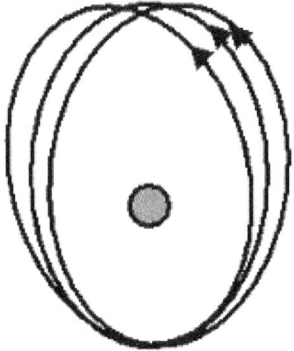

Figure 12-5 A planet in orbit around a sun with a marked precession of the ellipse.

All of the planets of our solar system precess, and that precession is almost entirely due to there being more than two bodies in the solar system. The additional planets cause most of the precession. There is a small amount due to the flattening of the Sun around its equator due to its own rotation. Initially, all of the planet's rates of precession were accounted for using Newtonian Mechanics, except for Mercury. In the case of Mercury, there is a 43 arcseconds per century difference between theory and observation. Later, other planets were shown to have small additional amounts of precession advance.

In order for this extra amount of advance to occur, there must be an additional force. This force causes Mercury to gain more velocity than expected when it nears the perihelion, where it is closest to the Sun, and lose more velocity than expected when it approaches the aphelion. In an elliptical orbit, a planet is always moving fastest at the perihelion, so what we are talking about is a very small extra change in velocity.

We have already shown that the mattermagnetic field pushes a planet toward the Sun due to the field generated by the Sun's rotation. This force was not accounted for in the original Newtonian computations, which led to the 43 arcseconds discrepancy. When the orbital motion is elliptical, this force is not usually directed toward the sun, but is everywhere tangential to the planet's direction of velocity. Because the acceleration due to the mattermagnetic force is not directed toward the Sun, it causes the planet Mercury to accelerate by a small amount versus the Newtonian model as it nears the perihelion. Conversely, this tangential force causes Mercury to slow down relative to the Newtonian model as it approaches its aphelion. This additional force is illustrated in Figure 12-6.

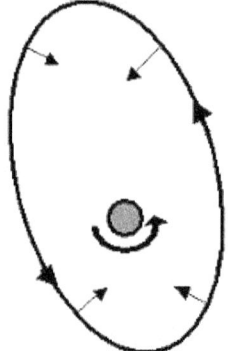

Figure 12-6 A planet in orbit around a rotating sun. The Lorentz force shown with small arrows is everywhere tangential to the planet's velocity vector and generally not directed toward the center of mass of the sun.

Note that since the force is directed first one way and then the opposite way as it rounds the perihelion, the larger part of this force is periodic in nature and does not lead to precession. There is, however, a residual, secular component that does cause precession. You can refer to Danby for the mathematical computations if you are interested.[68] Danby was able to computationally model the precession of the perihelion of Mercury by assuming a small modification to Newtonian Gravity. All he was lacking was an understanding of the nature of this small additional force. This modification to Newtonian Gravity is due to a mattermagnetic Lorentz Force, and it is this small additional force that is responsible for the classically unaccounted for 43 arcseconds of perihelion advance per century.

There is a somewhat related paper by Tom Van Flandern that may also be of interest to anyone who wants to delve deeper into the problem.[69] In that paper, he derives the additional perihelion advance by invoking a "tangential pseudo-force directed along the velocity vector." His pseudo-force is said to be due to variations in the light-carrying medium around the Sun, which leads to a Lorentzian velocity transformation. Instead, the actual cause is a real tangential force directed along the velocity vector. From that point on, the basic math works out the same.

In order for Einstein to come up with the same answer, he added four parts time dilation, subtracted two parts space contraction, and added one part relativistic mass increase. If you have two models and one requires a space contraction, time dilation recipe and the other does not, which one is most likely to be fundamentally correct?

Spiral Galaxies
Now we can turn to the largest flywheel of all. This one happens to not be solid, so it adds many additional degrees of freedom. This is perhaps the most substantial example of the mattermagnetic force among things that are astronomically observable. The spiral formation of a galaxy can simply not be accounted for by Newtonian gravitational theory, and explanations within the scope of General Relativity have not been satisfactory either. In a strictly Newtonian world, if a galaxy had arms they would be straight. In a strictly Newtonian universe, spiral galaxies would instead be disc galaxies with no arms.

The odd thing about the spirals is that stars that are farther away and, thus, traveling faster; are more strongly attracted to the center of the galaxy. Gravitational forces do not get stronger as distance increases; they get much weaker. What kind of force gets stronger with distance and the consequent higher tangential velocities? That would be a force that is the result of a vector cross product, a Lorentz force, not gravity. If we apply the mattermagnetic theory, the stars in the galactic core produce a magnetic field that stretches further out, just like our Sun does in our solar system. The stars farther from the center of the core have an even faster radial velocity. These stars, while not making up a large percentage of the total mass of the galaxy, cause most of the mattermagnetic field within the galaxy.

We can recall an equation from Chapter 8 shown below as Equation 12-5. In the case of a galaxy, we can see that the total mattermagnetic field depends on the mass and velocity of each point, each star (**dm**).

Equation 12-5

$$B_m = \frac{\mu_0 m}{4\pi} \int \frac{v \times \hat{r}}{r^2} dm$$

To look at it another, simpler way, we can consider a galaxy as a very large disc. The magnetic field is then proportional to the angular momentum **L**, which is equal to the moment of inertia **I** multiplied by the angular velocity ω, **L = Iω**. The moment of inertia **I** is on the other hand equal to half the total matter charge **m** multiplied by the radius **r** squared for a uniform disc, **I** = ½**mr²**. Figure 12-7 below shows such a disc. While the galaxy is nowhere near a uniform disc, it is easy to see that the mattermagnetic field increases rapidly when there are stars far from the center of the galaxy moving very rapidly.

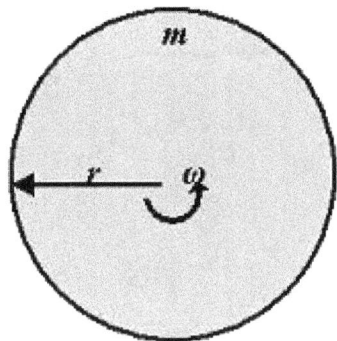

Figure 12-7 A rotating disc of matter charge m, radius r and angular velocity ω

Stars that are farther from the galactic core are drawn inward by the magnetic field, just as a planet is drawn inward by the mattermagnetic field of a rotating star. But in this case, the force is much stronger. These outer stars are also traveling at very great velocities, and since Matter Forces are proportional to velocity per the Lorentz equation $F_m = m_1 (u \times B_m)$, this force becomes very large. This force is so large that it exceeds the Newtonian gravitational force. There is no missing matter problem with respect to spiral galaxies. All along it has been a missing force problem.

The fast moving stars in the outer bands of the galaxy each produce their own strong mattermagnetic fields, and because they are more or less moving parallel to one another, the mattermagnetic field produced by one is directed opposite to the magnetic field produced by the other in the space in between. The situation is analogous to having a pair of parallel conductors, with their current moving in the same direction. Like the parallel conductors, the stars are attracted to one another. This magnetic attraction is greater than would be the case due to Newtonian Gravity alone. It is this mutual attraction between the fast moving outer stars that causes the spiral arms to form.

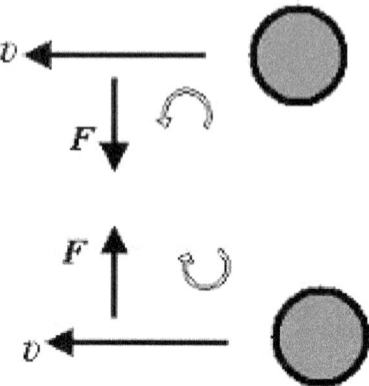

Figure 12-8 Two stars moving in the same direction produce a mattermagnetic field that attracts them toward each other.

Figure 12-8 shows two stars moving with the same velocity **υ**. They produce equal but opposite mattermagnetic fields in the space between them, as indicated by the curling arrows. This produces forces **F** drawing the two stars closer together.

Perhaps the best attempt at simulating the shape of spiral galaxies mathematically was accomplished by Anthony Peratt.[70] His work was based on the idea that galaxies were formed in a plasma state moving in response to electromagnetic forces. The images he developed are startlingly accurate at modeling each spiral galaxy subtype. But, while the plasma theory he favored may be dominant in interstellar medium and nebulae, stars are electrically neutral as observed from any substantial distance. His plasma-based theory was justifiably not taken very seriously. What was unfortunately missed is that Peratt succeeded in showing that spiral galaxies are formed in accordance with mathematics equivalent to Maxwell's Equations. It is just that it is not due to electromagnetic forces, but rather, due to mattermagnetic forces. Peratt's images are definitely worth a look.

Anomalous Acceleration in our Solar System

In our solar system, the planets and the asteroid belt also move at reasonably high velocities producing their own mattermagnetic fields. These fields in turn produce an additional inward directed force that causes satellites to be slightly more attracted toward the sun than would be the case due to Newtonian Gravity alone. This "anomalous, weak long-range acceleration" has been well documented by Anderson and his team. [71] They found evidence of this anomalous acceleration while examining data from the Pioneer 10/11, Galileo, and Ulysses space probes. The news generated a great deal of excitement since it was counter to Newtonian Gravity and General Relativity. It was concrete evidence that those two theories were incorrect. Even so, most mainstream physicists keep their fingers firmly in the ears and make noise while trying to ignore the well-documented facts that their theories for gravity are no longer adequate.

It is clear from Anderson's reports that the same magnetic component of the Matter Force responsible for spiral galaxy formation has been seen and measured within our own solar system. It should be pointed out, however, that part of the Matter Force has already been included in the gravitational constant G and the masses of the planets. What Anderson's team observed is only the difference between what had already been papered over and a somewhat greater magnitude force. It was the remainder that could not be easily covered up, which was discovered.

Conclusion

The physics of astronomical bodies is littered with examples where the forces of the Standard Model fail to adequately describe motion. Synchronous rotation, the precession of the perihelion of Mercury, the arms of spiral galaxies, and the anomalous acceleration in the Solar System are all concrete examples where the present theories have either failed outright or been suspect. Even the missing matter problem at the galactic level can be swept away, as the mattermagnetic force is far stronger than gravity in the arms of a spiral galaxy. When viewed collectively, it is easily seen that the Matter Force is consistently present across all ranges of scales of size from tops to galaxies. In this chapter we added the following to our list.

35) Synchronous rotation and tidal forces are due to a mattermagnetic force between rotating bodies of matter
36) The error in the precession of the perihelion of Mercury is due to the mattermagnetic force on Mercury due to the rotation of the Sun
37) The formation of spiral galaxies is due to the mattermagnetic force
38) The anomalous acceleration in our solar system is due to the mattermagnetic force

[66] J..H..Jeans, Astronomy and Cosmogony, Cambridge University Press, Cambridge p 352, 1928.

[67] A.G. Riess, et al, Observational Evidence from Supernovae for an Accelerating Universe and a Cosmological Constant, The Astronomical Journal, **116**:1009E1038, 1998 September.

[68] J. Danby. Fundamentals of Celestial Mechanics. Willmann-Bell Inc., Richmond, second edition, 1988.

[69] T. Van Flandern, "The Perihelion Advance Formula" Meta Research Bulletin 8:1-9 & 24-30, 1999.

[70] A. L. Peratt, "Simulating spiral galaxies", Sky and Telescope (ISSN 0037-6604), vol. **68**, Aug. 1984, p. 118-122.

[71] J. D. Anderson et al, Indication From Pioneer 10/11, Galileo, and Ulysses Data for an Anomalous, Weak Long-Range Acceleration, Phys Rev Letters 81:2858-2861, 1998.

Chapter 13: Solar and Planetary Dynamos

I propose now to examine magnetic phenomena from a mechanical point of view, and to determine what tensions in, and motions of, a medium are capable of producing the mechanical phenomena observed.[72]

James Clerk Maxwell, 1861

The Origin of Magnetic Fields

Another one of the great puzzles in physics is the source of the magnetic fields in stars and planets. What is the dynamo that produces the magnetic field? How does the dynamo come to exist? We can fall back on a field of study called magnetohydrodynamics, the study of the interactions between conductive fluids and magnetic fields. It is commonly accepted that the magnetic fields are caused by convection, where the fluids in the outer part of the body rotate, and by doing so produce a magnetic field.

The more fundamental question is, what causes this convection? Without this basic question answered it, is nearly impossible to develop reasonable models for the generation of magnetism and other phenomena related to the convection, such as sunspot activity on the Sun. The most likely answer based on the standard physics models is differential rotation, where the inner core of a body rotates faster than the outer layers. Such a velocity differential would be expected in a gas and plasma body like a star and a little less so in a body of molten rock. Rock can cool and solidify over time, which stops the dynamo at some point and makes the magnetic field largely disappear, except for residual ferromagnetic effects. In order for the Earth's dynamo to keep working more or less indefinitely, there must be a constant input of energy from some unknown source.

Matter Force Induced Convection

As with so many of the things we have considered before, in the context of the Matter Force, understanding the origin of the convection in stars and planets becomes trivial. When we turn our exploration inward and look at what happens to a non-solid rotating sphere, we see in the simplest first approximation that the mass can be divided in two. Half of the mass is in an inner sphere, and the other half is in an outer spherical shell. If the density is uniform throughout, the dividing line between these two halves is at a radius that is approximately 80% of the radius of the sphere.

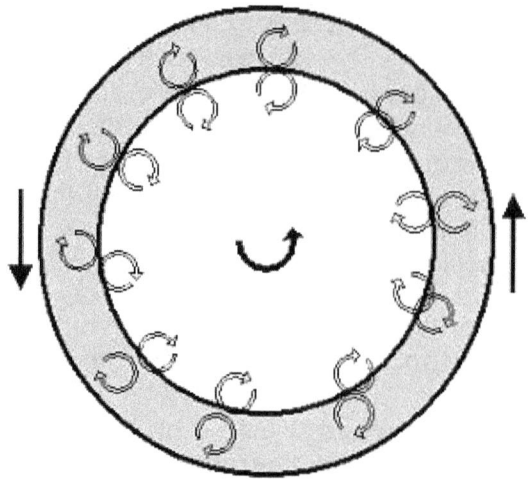

Figure 13-1 A body as seen from above rotating counterclockwise as shown by the black arrows with a spherical shell dividing the two halves of its mass. On either side of the spherical shell dividing the two masses there are eddy currents as shown by the circling arrows.

As the inner 50% of the mass rotates, it produces a counter-rotating mattermagnetic field outside of it in the volume outside the 0.8R radius. At the same time the other 50% of the total mass in the outer shell produces a counter-rotating mattermagnetic field in the region inside the 0.8R radius. These mattermagnetic fields produce two separate counter-rotating mattermagnetic loops or eddies that meet near the 0.8R spherical shell dividing the two fractional masses. This phenomenon is illustrated in Figure 13-1.

The mattermagnetic eddies cause real stirring of the matter, convection currents, in the region around the spherical shell dividing the two masses. This effect provides enough energy to melt rock and keep it molten as long as the body keeps rotating. Because there is a relatively small percentage of the total mass near the core, it is not as strongly affected and remains solid, in cases such as the Earth. Only those regions subject to the strongest eddy currents are molten. Except for the melting and solidification issues, the effect is the same in a gas and plasma body such as the Sun. Convection currents that are directly caused by the rotation of the Sun stir up the gas and plasma in the Sun's outermost regions.

The plasma of the Sun is conductive, as is the molten iron in the Earth's mantle, such that in both cases we have convective currents of conductive fluid. This leads us back to magnetohydrodynamic fundamentals, which yield the magnetic fields. Electrons find it easier to flow in the convection current loops than protons causing little electromagnetic loops to form, thus an electrically induced magnetic field forms.

It is also important to note that the convection current loops in turn cause their own counter current loops, rotating in the same directing as the body. These current loops also produce magnetic fields that oppose the first ones. In this way, the magnetic field of the body as a whole is a superposition of numerous small fields. This means that the magnetic field can vary over time and even change direction as the magnetic field changes due to somewhat periodic, yet somewhat random changes in fluid dynamics.

That said, the magnetic fields are still known to be stable for very long periods of time from a human perspective. Cycles such as the sunspot cycle are connected to a cyclical fluid dynamic process initiated by the mattermagnetic eddy currents. Previously, in the absence of a fundamental mechanism to be held responsible for the convection currents, it was difficult to explain the meta-stability of the magnetic fields. Without a primary driving force, it is hard to see how the convection currents could be anything but random. Now that we know that there is a primary driving force behind the convection currents tied to the angular velocity of a rotating body, we now have the basic tools to better understand the magnetic fields.

Conclusion

While it has long been widely accepted that the Sun's and Earth's magnetic fields are due to current loops due to fluid convection in their outer regions, there has not been a widely accepted force theory that describes the driving mechanism. Rotational differentials alone were not adequate to explain the convection currents and were certainly insufficient to sustain the melting in the Earth's mantle over time. Now with the inclusion of the Matter Force Theory, the picture becomes clear, and the driving mechanism can now be readily understood. With that now in place, over time our understanding of solar and planetary dynamo systems will improve greatly. In the meantime, by being capable of describing a driving force behind the convection currents in astronomical bodies, the Matter Force Theory has once again proved its value.

39) Convection and the dynamo effect in non-solid rotating bodies are due to the mattermagnetic force

[72] J.C. Maxwell, "On Physical Lines of Force" Part I "The Theory of Molecular Vortices applied to Magnetic Phenomena" Philosophical Magazine XXI, 1861. pp 161-175.

Chapter 14: The Matter Force

> *I have shown how the forces acting between magnets, electric currents, and matter capable of magnetic induction may be accounted for on the hypothesis of the magnetic field being occupied with innumerable vortices of revolving matter, their axes coinciding with the magnetic force at every point of the field. The centrifugal force of these vortices produces pressures distributed in such a way that the final effect is a force identical in direction and magnitude with that which we observe.*[73]
>
> James Clerk Maxwell, 1862

Matter Force Math

We have analyzed the behavior of the matter forces with respect to the most important large-scale interactions. It has been shown that matter interacts with matter in a manner comparable to Coulomb's Law, where matter is repelled by matte. This gives us a fundamental force to explain the known accelerating expansion of the universe, the so-called Dark Energy. As we have seen previously, it can be expressed as shown in Equation 14-1 where F_m is the resultant force between two bodies with matter charge m_1 and m_2 respectively at a distance r apart. The first term is a constant where 4π is related to the surface area of a sphere and ε_{0m} is the permittivity of the vacuum with respect to matter.

Equation 14-1

$$F_m = \frac{1}{4\pi\varepsilon_{0m}} \frac{m_1 m_2}{r^2}$$

One of the other things to jump out at us is that in this force model, matter has to be attracted to antimatter. So if both matter and antimatter occupy the same space, they will annihilate each other until nothing, or an excess of one or the other is left. This is a nice and tidy way to explain away another of the great mysteries of science, the question of why is there only matter in local space. It still leaves us with the question of why is there an excess of matter.

Matter charge attraction and repulsion can be more generally described by a law analogous to Gauss's Law which can be expressed in integral form as shown in Equation 14-2. The matter field M is determined by an integral of area A of a surface around a matter charge m. This is equivalent to the first of Maxwell's laws of electricity and magnetism.

Equation 14-2

$$\oint M \cdot dA = \frac{m}{\varepsilon_{0m}}$$

The next equation is equivalent to Gauss's Law of Magnetism as shown in Equation 14-3. This is a statement that the integral of the area A of a surface around a mattermagnetic source will always equal zero. This is equivalent to the statement that there are no mattermagnetic monopoles. All mattermagnetic phenomena are due to the rotation of dipoles, and those dipoles produce a north mattermagnetic pole in one direction and a south mattermagnetic pole in the opposite direction. The net mattermagnetic field when the intensity of the field at one pole is subtracted from the other is always zero. This is equivalent to the second of Maxwell's laws of electricity and magnetism.

Equation 14-3

$$\oint B_m \cdot dA = 0$$

The third equation, Equation 14-4, is analogous to Faraday's Law of Induction. It states that a changing mattermagnetic field $d\Phi m/dt$ produces a matter field M.

This is equivalent to the third of Maxwell's laws of electricity and magnetism.

Equation 14-4

$$\oint M \cdot dl = -\frac{d\Phi_m}{dt}$$

Faraday's Law is more practically used to compute electromotive forces, voltages, in a current-carrying conductor. This does not translate well to day-to-day uses involving matter, since matter is seldom viewed from the perspective of a current in a conductor, and when it is, the pipe or tubing carrying the matter is usually not being moved across a significant mattermagnetic field. The more useful form, which is a subset of Faraday's Law that applies to individual charges, is the Lorentz Force Law. The matter version of the Lorentz Force Law is shown in Equation 14-5, where *m* is the matter charge, *M* the matter field, *v* the velocity, and $\mathbf{B_m}$ the mattermagnetic field.

Equation 14-5

$$F_m = m(M + v \times B_m)$$

The last of the standard four equations, Equation 14-6, is based on Maxwell's version of Ampere's Law, which includes Maxwell's displacement current. This law basically says that the mattermagnetic field $\mathbf{B_m}$ is proportional to the matter current $\mathbf{I_m}$, along with a correction factor if the current loop, such as piping, cuts across a changing matter charge field $\boldsymbol{d\Psi_m/dt}$. This is equivalent to the fourth of Maxwell's laws of electricity and magnetism.

Equation 14-6

$$\oint B_m \cdot dl = \mu_{mo}\left(I_m + \epsilon_{mo}\frac{d\Psi_m}{dt}\right)$$

Of course this form of the equation is not at all useful for practical problem solving of matter forces. As pointed out previously, the momentum in Ampere's equation is taken into account with the current I without directly using a term for velocity. In the Matter Force case, we must always take the velocity into account. When discussing matter forces, it is far more useful to use a form that relates more directly to the Biot-Savart Law. The equation has been shown several times previously, but is shown again as Equation 14-7.

Equation 14-7

$$B_m = \frac{\mu_0 m}{4\pi} \int \frac{v \times \hat{r}}{r^2} dm$$

As a way of comparison, the Biot-Savart Law can be expressed as shown in Equation 14-8 with a line integral of the current I.

Equation 14-8

$$B = \frac{\mu_0}{4\pi} \int \frac{I \times \hat{r}}{r^2} dl$$

However, the Biot-Savart Law for a point charge, Equation 14-9, more closely resembles the matter charge form, since it incorporates a velocity term. This equation is due to Oliver Heaviside. The movement of charges q at velocities v generates the magnetic field at some point described by r.

Equation 14-9

$$B = \frac{\mu_0}{4\pi} \int \frac{v \times \hat{r}}{r^2} dq$$

All these equations can be expressed in differential forms and in other integral forms that are standard practice with Maxwell's equations. The simplest forms are used here to make it easier for readers less familiar

148

with more advanced mathematical forms. As it stands, these are sufficient to show that the mattermagnetic force laws are analogous to their electromagnetic force law counterparts. The mattermagnetic force can be expressed by a set of equations completely analogous to Maxwell's equations. In more practical usage, they can be expressed in forms equivalent to Coulomb's Law, the Lorentz Force Law, and the Biot-Savart Law.

I have shown how the forces acting between bodies of matter may be accounted for on the hypothesis of the polar and magnetic matter-antimatter fields consisting of innumerable zeptons either polarized or revolving about their axes coinciding with the mattermagnetic and matter charge force at every point in the field. The force exerted by these zeptons produces pressures distributed in such a way that the final effect is a force identical in direction and magnitude with that which we observe.

40) The Matter Force can be described in terms of Maxwell's equations

Note: For historical reference these equations were first published in my book *The New Physics* in 2001.[74]

---

[73] J.C. Maxwell, "On Physical Lines of Force" Part III "The Theory of Molecular Vortices Applied to Statical Electricity" Philosophical Magazine XXIII, 1862, pp 12-24.

[74] R. Fleming, The New Physics, self-published 2001 (rayfleming.com)

Chapter 15: The Electro-Matter Force

Thus, comparing a single moving particle of matter with a similarly-moving electric charge, describe a sphere round each. Let the direction of motion be the axis, the positive pole being at the forward end. Then in the electrical case the magnetic force follows the lines of latitude with positive rotation about the axis, and the flux of energy coincides with the lines of longitude from the negative pole to the positive. But in the gravitational case, although h still follows the lines of latitude positively, yet since the radial e is directed to instead of from the centre, the flux of energy is along the lines of longitude from the positive pole to the negative. This reversal arises from all matter being alike and attractive, whereas like electrifications repel one another.[75]

Oliver Heaviside, 1893

Combining the Two Principle Forces

Oliver Heaviside is quoted here because Heaviside was the first to attempt to combine the electromechanical with Newtonian mechanics. He saw gravity as possibly being an analogous force to electromagnetism, represented by analogs of the same four equations, forming a field theory now called gravitomagnetics.[75] He unfortunately had no way to know at the time that there indeed is a force causing matter to move away from matter, which would mean that matter is repelled from matter, and gravity is something a bit different. Heaviside, a self-taught and somewhat out of the mainstream physicist, recognized that Maxwell's 20 original equations could be represented more generally by only four equations. So he is one of the fathers of the modern four-equation set we call Maxwell's equations.

With the mattermagnetic force, we now have a force that is completely analogous to electromagnetism. The equations for each are identical in form and function to their electromagnetic counterpart. The interactions of both are due to interactions with zepton dipoles of the same type, the vacuum fluctuations of the zero-point field. Zeptons are responsible for the transmission of both types of forces.

It's important to note that each zepton has both types of charge and interact with both the electromagnetic and mattermagnetic fields. In order to be strictly correct it is necessary to combine these two forces into a single set of unified field equations. To start with the basics, we should look at Coulomb's Law for each as shown in Equations 15-1 and 15-2.

Equation 15-1

$$F = \frac{1}{4\pi\varepsilon_0}\frac{q_1 q_2}{r^2}$$

Equation 15-2

$$F = \frac{1}{4\pi\varepsilon_{0m}}\frac{m_1 m_2}{r^2}$$

First, we can recognize that a force is a force regardless of what causes it; so there is no reason to use a subscript e or m to distinguish the two. The charges q and m are different since one is electrical and the other matter. The distance r is the same, as is 4π. The remaining constant, the permittivity, is something that we need to figure out. The most important distinction we can see right away is that ε_0 has to be in units of electrical charge while ε_{0m} has to be in units of matter charge. The units most commonly used for ε_0 when solving a problem with Coulomb's Law is C^2/Nm^2 (Coulomb squared per Newton meter squared), where the Coulomb C is the unit of charge. Solving the equation leaves a result in Newtons N, the unit of force.

Matter Charge, Permittivity, and Permeability

Matter charge is not the same as mass. It is an intrinsic property of particles. As we will see, mass is the energy of the particle, something else entirely, even though the two have been confused for a long time - and not just linguistically. We could probably use mass units for matter charge in some computations and they would work out OK, but to be rigorously correct, they need to be analogous to electric charge units, preferably numerically identical. To this end it is best to numerically set the matter charge of an electron e_m equal to the electric charge e. Then in order to have more useful units, we need a unit equivalent to the Coulomb. I have taken to calling this unit the **Matt**, short for matter, of course, where **1 Matt = 6.24150965 x 10^{18} e_m**. Perhaps it is not the best name since the meter already has the abbreviation m, but it will do as a start. The Faraday unit of charge based on Avogadro's Number might be a more convenient, everyday unit, but consistency with electromagnetic convention is best in the long run.

If we look at the electrical permittivity of the vacuum ε_0, it has a value of **8.85 x 10^{-12} F/m** (Farads per meter) or **C^2/Nm^2**. The Farad is the unit of capacitance, so this constant is a statement of the vacuum's capacity to store electric charge. The permeability of the vacuum μ_0 is equal to **4π x 10^{-7} H/m** (Henries per meter), or approximately **1.257 x 10^{-6} H/m**. One of the more important relations in science is that the speed of light is equal to the inverse of the square root of the sum of ε_0 and μ_0. The relationship is shown in Equation 15-3. The same relationship must also be true for ε_{0m} and μ_{0m}, as shown in Equation 15-4.

Equation 15-3

$$c = \frac{1}{\sqrt{\varepsilon_0 \mu_0}}$$

Equation 15-4

$$c = \frac{1}{\sqrt{\varepsilon_{0m}\mu_{0m}}}$$

As with matter charge, it is best to retain the magnitude of permittivity and permeability for the matter equivalents while using units that are appropriate to discussions of matter. So, we can set the magnitude of ε_{0m} equal to ε_0 and the magnitude of μ_{0m} equal to μ_0. For now, the subscript m will be used to designate permittivity and permeability with regards to matter forces as well as in their units, as it will ultimately take a convention of scientists to sort out the issues and decide which units are most favorable.

The Electro-Matter Force

As before, we can come up with the four basic equations (Equations 15-5, 15-6, 15-7, and 15-8), which are completely analogous to Maxwell's equations. They are shown below in integral form. $\mathfrak{F}$ is used to refer to the combined electric and matter fields, and the symbol B refers to both the electromagnetic and mattermagnetic fields. The e and m subscripts are used to differentiate the two different fluxes. It may not be possible to combine the fields in any practical way when performing computations, so at one level these equations are symbolic of the unification of the two forces.

Equation 15-5

$$\oint \mathfrak{F} \cdot dA = \frac{q}{\varepsilon_0} + \frac{m}{\varepsilon_{0m}}$$

Equation 15-6

$$\oint B \cdot dA = 0$$

Equation 15-7

$$\oint \mathfrak{F} \cdot dl = -(\frac{d\Phi_{Be}}{dt} + \frac{d\Phi_{Bm}}{dt})$$

Equation 15-8

$$\oint B \cdot dl = \mu_e(I_e + \varepsilon_0\frac{d\Psi_e}{dt})$$

$$+\mu_{0m}(I_m + \varepsilon_{0m}\frac{d\Psi_m}{dt})$$

It may be more useful to combine some of the force laws as they can be more useful for direct problem solving - say for example, the combined Lorentz Force Law in Equation 15-9. As used previously, *q* and *m* are the electric and matter charge respectively, *E* and *M* are the electric field and matter field, and the magnetic field due to each is indicated by the appropriate subscript. In this case, the total force on a body can be determined by summing the two forces due to the different charges.

Equation 15-9

$$F = q(E + v \times B_e)$$
$$+ m(M + v \times B_m)$$

Conclusion
It has been shown that there is a preponderance of physical evidence to support the existence of a magnetic-like force occurring between electrically neutral bodies of matter that cannot be fully explained by Newtonian laws of physics, in particular his First Law of Inertia and his Third Law requiring an equal but opposite reaction. Additionally, this force can be used to explain phenomena from inertia to Dark Energy to the shape and form of spiral galaxies, which are not well

explained within the scope of the Standard Model. I have shown that the mattermagnetic force is computationally identical to the electromagnetic force, and that ultimately these two should be unified into a single force that governs the motion of matter, the Electro-Matter Force.

41) The Matter Force and electromagnetic force can be combined into a single force, the Electro-Matter Force

Note: For historical reference, a slight variation of these equations was first published in my book *The New Physics* in 2001.[76]

75 O. Heaviside, "A gravitational and Electromagnetic Analogy" Part 1, The Electrician **31** 281-282 1893

[76] R. Fleming, The New Physics, self-published 2001 (rayfleming.com)

Chapter 16: Mass

Matter uses and stores energy as inertia, just like a steam engine that uses the energy in steam and stores energy in inertia as potential energy [...] All components of a body are animated by infinitesimal but rapid movements equal to perhaps the vibration of the ether. It must be concluded that the matter in any body contains the sum of the energy represented by the entire mass of that body if it could move through space with the speed of a single particle. [...] The matter of any body contains within it a sum of energy represented by the entire mass of the body [...] Nobody will easily admit that, stored in a latent state, in any kilogram of matter, completely hidden to all our investigations, hides such a sum of energy, equivalent to the amount that can be extracted from millions and millions of kilograms of coal. [77]

Olinto de Pretto, 1903

Proton and Electron Mass

Two of the great mysteries of physics are the origin of mass, and the mysterious mass ratio between the proton and electron of **~1836**. There have been speculations about the derivation of mass as energy for more than a century, ever since Einstein popularized the equivalence $E=mc^2$. Indeed de Pretto, quoted above, preceded Einstein in recognizing the equation for the mass-energy relationship and perhaps more importantly recognized that "all components of a body are animated by infinitesimal but rapid movements equal to perhaps the vibration of the ether." While Einstein went off on his own special tangent, de Pretto had pointed the way for us to understand the true nature of mass.

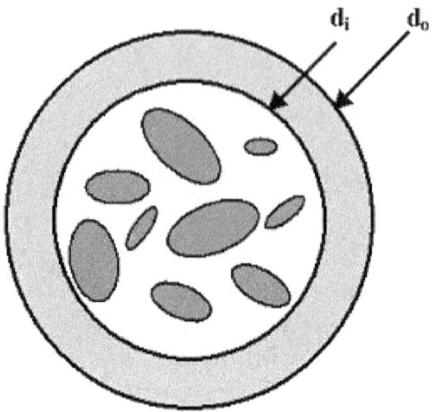

Figure 16-1 An illustration of the particle shell with elliptically shaped vacuum fluctuations in the middle and the spheres describing the inner and outer shell diameters indicated.

Revisiting the Casimir Effect, we can start with the assumption that a particle such as a proton approximates a spherical shell, which can be treated as a Casimir cavity. Such a cavity is shown in Figure 16-1. Wavelengths of vacuum fluctuations shorter than the inner diameter d_i of the shell exist inside it. Vacuum fluctuations with wavelengths larger than the outer diameter d_o of the shell exist outside it. The vacuum fluctuations with wavelengths between the inner diameter and outer diameter of a spherical shell are excluded. Based on this idealized model, it is obvious that an attempt should be made to compute the vacuum energy excluded by a spherical shell the diameter of a proton, and compare that to the proton's mass-energy.

Vacuum Energy of a Proton
The spherical shell simulating a proton was taken to have a radius equivalent to the proton charge radius as published in CODATA 2010 **0.8775(51) x 10⁻¹⁵ m** (femtometers = fm). Likewise the CODATA 2010 value for the mass-energy **938.272046(21) MeV** was used. From those values the mass-energy density of the proton was computed to be **3.315 x 10³⁸ GeV/cm³**.

The vacuum energy excluded by a spherical shell can be computed using Equation 16-1.[78] The angular frequencies ω_1 and ω_2 are related to the outer diameter d_o and inner diameter d_i respectively, where $\omega_1 = 2\pi c/d_o$ and $\omega_2 = 2\pi c/d_i$. The speed of light is designated as c and $\hbar$ is the reduced Planck's constant.

Equation 16-1

$$\rho = \frac{\hbar(\omega_2^4 - \omega_1^4)}{8\pi^2 c^3}$$

To get a first approximation, we can initially ignore the ω_1 term and compute the energy density for all wavelengths from the proton diameter, **1.755 fm** and larger. This initial result is **4.106 x 10^38 GeV/cm³**. Right away we can see that the excluded vacuum energy of a proton-sized spherical shell is a good approximation of the mass-energy of the proton. If we then set the energy density ρ equivalent to the mass-energy density of the proton and solve for ω_1, we find a value for d_o of **2.649 fm**.

If we instead assume that the measured diameter of the proton based on the charge radius is the average of the inner and outer diameters and set the vacuum energy density equal to the known mass-energy density of the proton, we can calculate those inner and outer diameters. In this case, d_i=**1.586 fm** and d_o=**1.924 fm**. The difference in diameters is **0.338 fm**, which equates to a shell thickness of **0.169 fm**.

Note that the exclusion of wavelengths on the order of the shell thickness was not considered. This is due to the idea that the proton shell must be a virtual structure composed of vacuum fluctuations or other virtual particles such as Feynman's original partons. This is necessary because of the spin of the proton and the speed of light limit, for if the proton were composed of stable matter, the outer shell would reach velocities greater than the speed of light. The logically simplest way around this difficulty is to assume that the proton shell is a virtual structure with pair production and annihilation events progressing sequentially around the circumference of the shell in a manner simulating high-velocity rotation. Additionally smaller, higher-energy vacuum fluctuations still exist in and around this structure, so they are not being excluded and, therefore, do not contribute to the mass-energy.

The value of the difference in diameter is interesting in that it is similar to the wavelength of a virtual proton-antiproton pair at the pair production energy, **0.330 fm** = $\boldsymbol{\lambda_{p\text{-}a}}$ = $\boldsymbol{hc/4m_p}$, where $\boldsymbol{m_p}$ is the mass energy of the proton. The shell thickness is essentially what we would expect if the shell were composed of vacuum fluctuations.

Based on these simple computations, it is readily apparent that the proton mass-energy at rest is equivalent to the energy of the excluded zepton wavelengths, if we assume that the proton is a virtual spherical shell. It is important to note that the solution for a given radius or shell thickness is unique, since the density of a sphere varies with the radius cubed and the vacuum energy density varies with the diameter to the forth power.

Vacuum Energy of an Electron

The electron is a bit more trouble as there are several radii associated with it, or possibly none at all for a so-called bare electron. In the minds of most physicists of today the point-like particle model is favored, but to quote MacGregor from *The Enigmatic Electron*: "a rather compelling case can be made for an opposing viewpoint: namely *that the electron is in fact a large particle which contains an embedded point-like charge.*"[79]

One of the minor travesties in physics is that in physicists' haste to conduct particle experiments at increasingly higher energies, an experimental value for the electron radius in the realm of Compton scattering

160

was never firmly established. What is known is that the scattering radius of the electron with respect to photons and other electrons is many orders of magnitude larger than the scattering radius due to protons and other high-energy particles.

Using Equation 16-1, if we allow the mass density of the electron to vary by trying various diameters while computing the vacuum energy for the same diameters, we quickly find that the two mass-energy density numbers coincide at or near the Compton wavelength. This is not terribly surprising, as the Compton wavelength is associated with the scattering of photons by an electron and with mass by the relationship $\lambda_c =$ $h/m_e c$, where m_e is the rest mass of the electron. In quantum electrodynamics, the Compton wavelength is related to the charge radius and equates to the rest mass-energy of an electron.

We can then compute the mass-energy density of the electron using the Compton wavelength as the electron diameter, CODATA 2010 value **2.4263102389(16) x 10-10 cm** (= picometers = pm), and the CODATA 2010 value for the electron's mass-energy **0.510998928(11) MeV**. The energy density of the electron is then calculated to be **6.833 x 10^25 GeV/cm^3**.

If we initially set d_i equal to the Compton wavelength, the energy density of the excluded wavelengths from that point and larger is **1.124 x 10^26 GeV/cm^3**, which is very close to the electron mass-energy density. Then, by setting the vacuum energy density equal to the mass-energy density of the electron, we can solve for $d_o =$ **3.066 pm**, which is only slightly larger than the Compton wavelength. This first approximation is in good agreement with the electron rest mass and gives a reasonable diameter relative to the Compton wavelength.

Then as before, we can set the average diameter of the electron shell equal to the Compton wavelength and solve for the inner and outer diameter. We find they are

d_i = **2.123 pm** and d_o = **2.472 pm**. The difference between those two diameters is **0.349 pm**, giving a shell thickness **0.175 pm**.

The electron Compton shell must likewise be a virtual structure so that it does not violate the speed of light limits and is transparent to shorter wavelength higher energy particles, such as a proton. The difference between the inner and outer diameter when the electron diameter is equal to the Compton wavelength, is a little over half the virtual electron-positron wavelength of **0.6066 pm** = $\lambda_{e\text{-}p}$ = $hc/4m_e$ at the pair production energy. If instead we set that difference equal to $\lambda_{e\text{-}p}$ and solve for the average radius, where the vacuum energy density equals the mass-energy density, we find that the average radius r = **1.592 pm**, and the inner and outer diameters are d_i = **2.880** and d_o = **3.487**. This leads to the possibility that the proton and electron can have identical shell structures made of proton-like virtual particles and electron-like virtual particles, respectively.

At this point it is clear that the masses of the proton and electron are due to the vacuum energy they exclude. The proton to electron mass ratio of 1836 is also accounted for in the process. The simplicity of the technique is compelling even though a major reevaluation of what we think we know about particles will be necessary to make this result consistent with a broader particle theory. Having a fundamentally electromagnetic description of mass brings up some additional questions. The biggest of which is that if mass is a purely electromagnetic phenomena then gravity must be as well. We will see in the next chapter that this is the case.

Particle Structure

In addition to his groundbreaking force model, Casimir proposed a semi-classical model for the electron based on the idea that if an electron were a spherical shell composed of an evenly distributed charge, then the Casimir Force on the shell may be able to oppose the Coulomb repulsion.[80] This theory has already been shown to be incorrect by Boyer, as the Casimir Force on a conducting infinitely thin spherical shell is directed outward.[81] It is not clear, however, if that will still be the case if the spherical shell is effectively transparent to shorter vacuum fluctuation wavelengths, not much smaller than the particle radius. After all, it is those shorter wavelengths that produce the outward directed force.

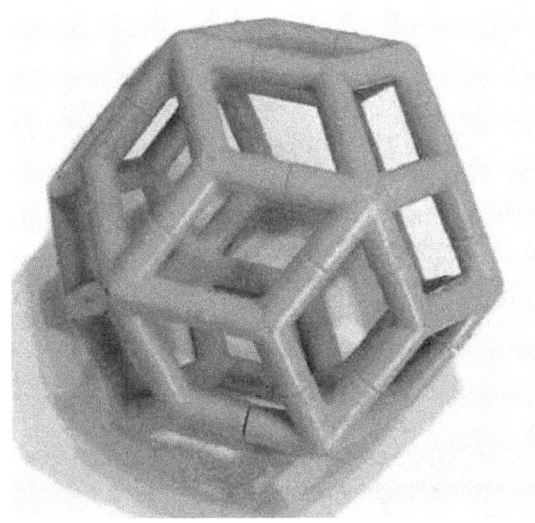

Figure 16-2 A rhombic triacontahedron. Could this be what a proton and electron look like? [Pictured with permission Roger von Oech's Star-Ball® www.CreativeWhack.com]

On the other hand, if an electron, for example, is composed at least partly of virtual electrons and positrons, then the shell could contain both positive and negative charges in an arrangement analogous to a fullerene molecule. This allows for the possibility that the Coulomb force could be neutralized geometrically. If we examine a rhombic triacontahedron as shown in Figure 16-2, we find it has 12 outer vertices and 20 inner vertices. If we assume the outer vertices are occupied by one type of charge and the 20 inner vertices the opposite charge, we can calculate the static Coulomb force on each charge with respect to the center of the polyhedron. What we find is the inward directed Coulomb force on the outer 12 charges combined is almost equal, ~99%, to the outward directed force on the 20 inner charges combined. The rhombic triacontahedron is almost stable with respect to Coulomb forces alone, so only a small addition force, such as a magnetic force, is required to make it electrically stable. If a particle can be electrically stable without any additional forces required, that will overcome one of the more serious objections to a semi-classical geometric particle model.

The rhombic triacontahedron only has a single edge length such that the energy between electrons and positrons could be identical between each vertex. It is also a self-dual such that it could have 12 positive outer charges and 20 inner negative charges or 12 negative outer charges and 20 positive inner charges and be stable in both cases. In this manner, two complementary stable particles such as an electron and positron could be based on the same geometric form while having different net charges. These two complementary forms would also annihilate each other if they occupied the same space, meeting a requirement of basic particle theory. Also, since the charge would be a property of the shell, the so-called bare electron may simply be the center of the electron with no properties attributable to it. A proton would likewise have nothing inside its shell.

These thoughts about a simple rhombic triacontahedron virtual shell structure are no doubt highly speculative and certainly may not pass the numerology test. But, if the proton and electron are not dimensionless, they will have to have some kind of structure, and any good theory begins with a little speculation. The important point to take away is that the proton and electron can have identical virtual structures composed of proton-like vacuum fluctuations and electron-like vacuum fluctuations, respectively, while their mass is accounted for as a simple electromagnetic effect, the exclusion of vacuum energy.

Proton and Electron Mass Difference

So far, the mass-energy difference between the proton and electron has not been addressed. In order to understand that difference we have to go back to electromagnetic theory. It was noticed previously that a negative electric charge moving through an electromagnetic field behaves in the same way that a positive charge would if it was moving backwards in time. This is identical in principle to the way that antimatter is equivalent to matter moving backwards in time per the standard conventions of Feynman diagrams. Due to Dirac's equation, we also have the convention that antimatter is a form of negative energy with matter is a form of positive energy. Similarly, negative electric charge can be thought of as form of negative energy.

This is not to say that you just place a minus sign in front of it and be done with it, but rather that the state of negative and positive energy is something more fundamental. Charge interactions between negative and positive charge-energies can be more properly thought of as opposite charge-energies, attempting to cancel each other in order to reach a zero-energy state, while like charge-energies tend to move away from each other in order to reduce the average energy of space. In either case, the charges are acting in such a way as to reduce the local energy of space.

Then if we consider a proton, we find it has both positive charge-energies opposite both negative charge-energies, such that they are strongly attracted. The electron on the other hand has positive and negative charge-energies on each end of the dipole, such that there is a much weaker attraction. This difference appears to be the most likely reason for the difference in mass-energy.

Conclusion

It has been shown that the mass of the proton is equal to the vacuum energy it displaces, if it were a spherical shell structure with an average radius equal to its charge radius. The electron mass can also be derived based on the vacuum energy it excludes, if it were a spherical shell with an average diameter equal to its Compton wavelength. Alternatively, the electron has a radius of ~1.6 pm if it has a structure that is geometrically equivalent to the protons, assuming both are proportional in size to their respective virtual particle wavelengths. Experimental verification of the electron charge radius will be necessary to determine the actual radius of the electron with respect to photon and electron scattering.

For many decades it has been written that de Pretto's work did not really have priority over Einstein's based on the date it was published, because Einstein derived $E=mc^2$ by using Special Relativity. It is, however, de Pretto who more properly tied this energy to the æther, or in modern terms the zero-point field. Perhaps its is de Pretto who deserves the acclaim.[82]

42) A particle's mass is equal to the vacuum energy displaced by the particle

[77] O. De Pretto, "Ipotesi dell'etere nella vita dell'universo (Hypothesis of Aether in the Life of the Universe)". "Reale Istituto Veneto di Scienze, Lettere ed Arti" (The Royal Veneto Institute of Science, Letters and Arts) LXIII (II): 439–500. (accepted November 23, 1903 and printed February 27, 1904)

[78] P.W. Milonni, The Quantum Vacuum, Academic Press LTD, London 1994, p. 49

[79] M.H. MacGregor, The Enigmatic Electron, Kluwer, Dordrecht, Netherlands 1992 p. 5

[80] H. B. G. Casimir, Introductory remarks on quantum electrodynamics, Physica 19, 846 (1953). DOI: 10.1016/S0031-8914(53)80095-9

[81] T.H. Boyer, Quantum Electromagnetic Zero-Point Energy of a Conducting Spherical Shell and the Casimir Model for a Charged Particle Phys. Rev. 174, 1764 (1968).

[82] J.H. Poincare also deserves credit for the mathematical expression since he set mass equal to E/c^2 in 1900. J. H.Poincaré, Arch. neerland. sci., 2, 5,232, 1900, J.H. Poincaré , In Boscha 1900:252

Chapter 17: Gravity

Gravity is produced, in my view, by Exceedence of the Speed of the Particles of this Matter which impinges on the Earth (for example, or some coarse Atom of which it is composed) over their Speed when they are reflected...[83]

Nicolas Fatio de Dullier, ~1690

Gravity in History

Early physicist-philosophers such as Descartes and da Vinci favored the idea that gravity could be explained mechanically that the medium of space pushed or pulled on astronomical bodies causing them to move. Descartes went further, stating that the medium was filled with vortices that rotate as a means of explaining orbital motion. The vortex concept is at its heart the mattermagnetic force theory, but Descartes' theory was unfortunately never developed as such. The reason is that after Newton published his *Principia*, Descartes' theories were largely set aside. Newton did an outstanding job discrediting the idea of large-scale vortices of the type envisioned by Descartes while not considering that the vacuum could be composed of a nearly infinite number of small vortices, dipoles that could accomplish what Descartes envisioned.

There was one of Newton's contemporaries who did make an attempt at a mechanical theory to explain gravity; it was a young Swiss scientist named Nicolas Fatio de Duillier.[83] He originated the push theory of gravity, also called the shadow theory of gravity. This theory was later popularized by Le Sage and often is referred to as Le Sage's theory of gravity, but credit is truly due to Fatio. Fatio's idea was that space is filled with corpuscles, and that as these corpuscles move through space, they randomly strike astronomical bodies, pushing them. If we then consider two bodies, the first body shadows the second body from the corpuscular pressure force coming from behind the first

167

body, and in this way there is more force pushing the bodies together than pushing them apart. The concept as illustrated by Le Sage is shown in Figure 17-1, while Figure 17-2 is an illustration showing the shadowing effect.

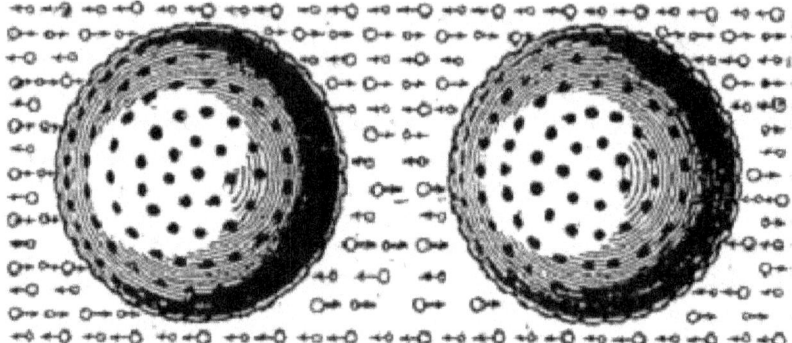

Figure 17-1 An illustration of the Fatio's Push Gravity Theory by Le Sage.

To this day, Fatio's Theory is the most compelling classical mechanical theory of gravity and many brilliant physicists have considered the theory, unfortunately to only reject it.[84,85,86,87] In principle, the push theory could describe the *1/r²* varying force required, but there were many problems. One of the first and most fundamental of the questions was, what is the nature of the corpuscles?

Star 1   **Star 2**

The "Shadow" on the Star's Surface

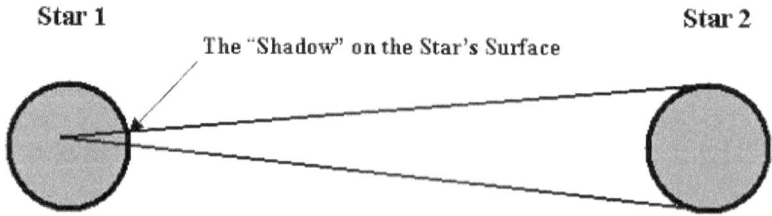

Figure 17-2 An illustration of the shadow from the shadow gravity part of the theory.

Lord Kelvin and others pointed out that a purely kinetic interaction between the corpuscles and astronomical bodies would generate heat. So much heat it turns out that the bodies would be vaporized.[84] Entering the last century, that is the way things stood. A mechanical theory of gravity was not deemed important, a mattermagnetic description of orbital motion had been long ignored, and the most likely mechanical description of gravity was widely discredited.

Solving the Problems with Fatio's Theory

The first and most important problem to us is what is the true nature of Fatio's corpuscles. If you have read this far, the answer is an easy one to figure out. Fatio's corpuscles are vacuum fluctuations, the zeptons that fill all of space. There is simply nothing else they could be.

One of the first problems addressed by Fatio was of the porosity of matter. If a body is perfectly solid and is pushed by a gas kinetically, the force on the body is proportional to the surface area, not the volume. The gas pushes uniformly across the surface. In order for the force to be proportional to mass, the interaction must involve all the body, throughout the entire volume, the entire mass. Fatio wrote that despite the apparent heaviness of gold, it is entirely possible that gold contains a million (10^6) to a trillion (10^{12}) times more void than substance. He based this idea on the fact that some materials are transparent to light. This was amazing insight into the composition of matter for its day. So amazing was his insight that it was not believed, and for the next 200 years, the impossibility of such an assertion was grounds to dismiss his theory out of hand. We of course now know that he was correct and, if anything, his estimate was too low by quite a few orders of magnitude. Based on our present knowledge, it is the idea that the force is proportionate to surface area that can be dismissed out of hand, particularly now that we know that mass is nothing more than excluded vacuum energy.

In order for the pressure force to be exactly proportional to mass, the porosity of matter needs to be effectively infinite. We know that the zeptons, which fill all space, can be many orders of magnitude smaller than even the smallest fundamental particle, making matter, for all intents, infinitely porous. The other part of this problem is that energy of the corpuscles needs to be infinitely energetic. Once again the energy of the zeptons is so much greater than the energy of matter that it is effectively infinite in energy. The problems of infinite porosity and infinite energy are easily met by considering that Fatio's corpuscles are zero-point vacuum fluctuations.

Another problem was the question of elastic or inelastic scattering. The Push Gravity Theory does not work if the interactions are entirely elastic. In that case, the corpuscles push the body and the body pushes on corpuscles on the other side, sending them along the initial direction with no loss of energy. No shadowing effect occurs if the interaction is completely elastic. Fatio went about deriving an equation to describe the forces on a gas that experiences only inelastic scattering, coming up with a formula that is essentially equivalent to Bernoulli's equation for inelastic media many years before Daniel Bernoulli published his own equation. Fatio and Jacob Bernoulli, Daniel's uncle, knew each other well, with Jacob possessing a copy of Fatio's paper. So one wonders how much of his nephew Daniel's ideas originated with Fatio? To continue, Fatio and others generally presumed there would be a high degree of inelasticity to the interactions, however that may come about.

One of the standing criticisms of the Push Gravity Theory is that if the energy were absorbed, producing heat, the amount of energy absorbed would instantly vaporize everything. Kelvin, Maxwell, and Poincaré pointed this out in their analysis of Fatio's theory.[84,85,86] It is true that if the energy due to the vacuum, the zero-point energy, were absorbed by regular matter, it would be vaporized instantaneously, and there would be no solid matter anywhere. Based on the fact that we are here, we know that the heat-producing model of the kinetic corpuscular interaction is not accurate, as long as the corpuscles are zero-point vacuum fluctuations.

Richard Feynman was the latest to point out another major objection: that the mechanism behind Fatio's Theory would produce drag and orbits would decay.[87] This result is obviously true in the case of corpuscles interacting kinetically with a body. However, once Fatio's corpuscles are replaced with zero-point vacuum fluctuations, the only drag effects are inertia and other Lorentz forces due to Matter Force interactions. Matter-based Lorentz forces are already accounted for in orbital mechanics, so there is no drag effect.

The Casimir Effect
The problems of inelastic scattering and heat should have been cleared up when Casimir and Polder published their paper on the Casimir Effect.[88] As previously discussed, Casimir hypothesized that an object is pushed on by vacuum fluctuations via a London-van der Waals Force mechanism. In the simplest case, two plates that are close together exclude certain wavelengths of vacuum fluctuations from existing in the space between them. As more and more wavelengths are excluded from the cavity, there is less force pushing the objects apart than pushing them together.

That sounds a lot like Fatio's Push Gravity Theory, doesn't it? Casimir recognized that the vacuum fluctuations press on matter by means of a London-van der Waals Force, an electric force between the dipoles of the vacuum. Motion between two bodies is then caused by a difference in pressure between the pressure pushing the bodies together and the pressure pushing the bodies apart. Casimir actually solved the biggest problems with Fatio's Theory. His paper showed that the corpuscles are vacuum fluctuations. The experimental evidence for the Casimir Effect tells us that the vacuum fluctuations do interact with bodies of matter inelastically. The experimental evidence also confirms that the interactions between vacuum fluctuations and bodies of matter do not generate excess heat or drag. The Casimir Effect theory in one striking blow overcomes almost all the principal problems that are said to exist with Fatio's Theory. Almost all the perceived problems with Fatio's Theory are tied to dealing with Fatio's corpuscles through a kinetic theory of gases type model, when the corpuscles are actually a sea of zeptons, the zero-point field. Zeptons in this case are not interacting purely kinetically, but rather through Electro-Matter Force interactions.

The Van der Waals Force
The hypothesis that vacuum fluctuations are the source of gravity does have a problem. The pressure produced by the zeptons of the zero-point field must come about due to something like a van der Waals Force. Unfortunately, van der Waals Forces in general do not meet the distance squared law ($1/r^2$) requirement of any potential theory of gravitation. Note the r^2 in the denominator of Newton's equation for gravity. The force due to the standard van der Waals type interactions for vacuum fluctuations varies due to distance, geometry, and a number of other factors from $1/r^3$ to $1/r^7$, but generally not $1/r^2$. Because of that, the force becomes insignificant at relatively short distances, too short to account for gravity.

The potential for a true breakthrough first occurred in 1967 when Andrei Sakharov, the well known Russian physicist and dissident, postulated that gravity was due to the "metrical elasticity" of space, which was generally presumed to be due to vacuum fluctuations.[89] This is sometimes viewed as something akin to Fatio's Theory; however, Sakharov was set on bending his theory to match the space-time curvature theory from General Relativity. He never mentioned Fatio or Le Sage in his paper and made no attempt to fit it directly into an older Push Gravity Theory. Sakharov did introduce the cutoff frequency problem, but as we now know from the Casimir Force model, forces due to the high-energy short wavelength vacuum fluctuations are balanced on both sides of a body, so there never really was a problem.

Hal Puthoff, another physicist who has been on the forefront of research into the science of zero-point energy, made a significant attempt to show how vacuum fluctuation *zitterbewegung* could be responsible for the gravitational force.[90] In his paper, Puthoff first successfully derives an energy density that is proportional to *$1/r^4$*, which is not unexpected. He continued on to mathematically massage that into a *$1/r^2$* varying force. The first part of his paper is generally accepted. The second half was met with skepticism by Steve Carlip who stated, "a paper by H. Puthoff, which claims to derive Newtonian gravity from stochastic electrodynamics, contains a serious computational error. When the calculation is corrected, the resulting force is shown to be non-gravitational and negligible."[91] Carlip's paper is unfortunately consistent with most scientists' view of Puthoff's paper, which leaves us with still having to convincingly explain how a *$1/r^2$* varying force can be derived from vacuum fluctuations.

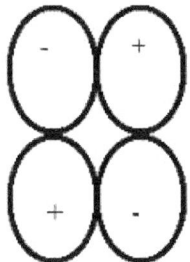

Figure 17-3 Two electric dipoles side by side with their charges canceling.

Adding the Matter Force to our understanding of van der Waals Forces changes everything. When we consider zepton dipoles with only positive and negative electric charge, there is a van der Waals Force. But at a distance, the dipoles cancel and changes in dipole moment become negligible, giving the appearance of neutral space as the force falls to zero. Two dipoles with positive and negative charges will align with their opposites together as shown in Figure 17-3. There is no way for them to form extending long-range fields that are necessary to cause a long-range $1/r^2$ varying force.

Table 17-1

Example Particle	Electric Charge	Matter Charge
Proton	Positive	Matter
Antiproton	Negative	Antimatter
Positron	Positive	Antimatter
Electron	Negative	Matter

When we include the Matter Force, we now have to consider a system with a combination of two different types of dipoles with combinations of two different types of charges. That gives us four different matter and electric charge combinations, as shown in Table 17-1.

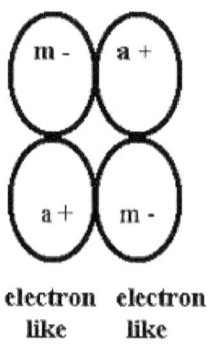

electron electron
like   like

Figure 17-4 Two electron-like zepton pairs side by side with both their electric and matter charge dipoles canceling

The electron-like zeptons when paired with other electron-like zeptons largely cancel, and those interactions lead to the standard short-ranged force, as illustrated in Figure 17-4. When proton-like zeptons

174

align adjacent to other proton-like zeptons, the charge forces also drop to zero after a fairly short difference. Once again, a short-ranged force is the result.

Things change when we look at an electron-like zepton adjacent to a proton-like zepton. If they lie next to each other with their positive and negative electric charges canceling, their positive and negative matter charges are in a repulsive configuration. And, if they lie next to each other with their positive and negative matter charges canceling, their positive and negative electric charges are in a repulsive configuration. This leads to basic Coulomb repulsion, a $1/r^2$ varying force, regardless of orientation. A basic zepton pair is illustrated in Figure 17-5.

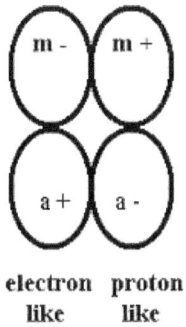

electron proton
  like   like

Figure 17-5 A zepton pair one electron-like and one proton-like adjacent to each of with their electric dipoles canceling and their matter dipoles repelling

In either case, whether the electric charges are canceling or the matter charges are canceling, the other charges in these pairs are repelling each other, due to Coulomb repulsion. A field of such zepton pairs forms a repulsive Coulomb pressure field, but a field that is electric and matter neutral. This field extends out infinitely just as we expect with any Coulomb field. This Coulomb repulsion does not fall off with distance over a short-range like a normal van der Waals Force does. Instead, the Coulomb repulsion component produces pressure that varies with the inverse of the distance squared ($1/r2$). It is the Coulomb forces between electron-like zeptons and proton-like zeptons that is the cause of gravity. This along with electro-matter charge distribution is also the underlying cause for electro-matter forces obeying the $1/r^2$ rule. If we think about it a bit, a base pressure of the vacuum that varies in accordance with the $1/r^2$ rule is necessary for all forces to vary in the a manner consistent with experiments and theory.

In a vacuum, this Coulomb field is evident as a pressure force, which is uniformly isotropic. In other words, it is the same strength in all directions in free space. Because this Coulomb field is due to virtual dipoles that behave as induced dipoles in a manner consistent with London dispersion forces, it can be thought of as a new type of long-range London-van der Waals Force. In light of that, it will generally be called the long-range van der Waals Force throughout the remainder of this book.

Having matter in the vicinity disrupts the force, reducing the long-range van der Waals pressure coming from the direction of the body of matter. The reduction in the force causes bodies to move together that are effectively shadowing each other, exactly as expected within the scope of the Fatio's Push Gravity Theory. The interaction is otherwise in accordance with Newtonian gravity in that it is essentially a point-to-point force between the centers of mass of the objects.

Long-range van der Waals Force-based gravity is, however, different from the Newtonian form in that the Newton's Gravitation Constant (G) has a different magnitude. As discussed earlier, the old **G** constant is actually a summation of the long-range van der Waals gravitational force, matter repulsion, and matter motion induced Lorentz forces. In the case where there is no orbital motion, Newtonian Gravity simplifies to the sum of the long-range van der Waals gravitation force and the matter repulsion force. We know from observation that the gravity is stronger than matter repulsion over most ranges of scale, at least up to the size of small galaxies. The absolute magnitude of both forces is stronger than Newtonian gravity. Depending on the velocity of the orbit, such as in spiral galaxies, the Lorentz forces can be much stronger than either of the other two forces.

At this point you might say that if this theory of gravitation behaves effectively the same as Newtonian gravity then it fails all the standard tests needed of a model proposed to replace General Relativity. In response, it could be said that General Relativity fails multiple tests of gravity including synchronous rotation and spiral galaxy formation, but that would be incorrect, since those are due to the Matter Force. Both theories will be put through a series of tests in a coming chapter after we examine relativity more closely.

In this chapter it has been shown that gravity is a long-range van der Waals Force due to the pressure produced by the natural motion of electron-like and proton-like zeptons. The manner in which the force operates is consistent with Fatio's Push Gravity Theory, which is similar in principle to the Casimir Effect.

> 43) Gravitational interactions act in a manner consistent with Fatio's Push Gravity Theory
> 44) Gravity is a long-range van der Waals Force

[83] N Fatio de Dullier, "De le Cause de la Pesanteur" (ca 1690), Edited version published by K. Bopp, Drei Untersuchungen zur Geschichte der Mathematik, Walter de Gruyter & Co. pg 19-26 1929.

[84] W. Thomson, (Lord Kelvin), "On the ultramundane corpuscles of Le Sage", Phil. Mag. 45: 321–332, 1873.

[85] J.C. Maxwell, "Atom", in none, Encyclopedia Britannica, 3 (9th ed.), pp. 38–47(1875).

[86] H. Poincaré, "The Theory of Lesage", The foundations of science (Science and Method, 1908), New York: Science Press, pp. 517–522, 1913.

[87] R.P. Feynman, The Character of Physical Law, The 1964 Messenger Lectures, pp. 37-39, 1967.

[88] H. B. G. Casimir, and D. Polder, "The Influence of Retardation on the London-van der Waals Forces", Phys. Rev. 73, 360-372, 1948.

[89] A.D. Sakharov, Vacuum Quantum Fluctuations in Curved Space and the Theory of Gravitation, Doklady Akademii Nauk SSSR Vol. 177 No. 1 Pg 70-71, 1967.

[90] H.E. Puthoff, Gravity as a Zero-Point-Fluctuation Force, Phys Rev A 39:2333, 1989.

[91] S. Carlip, Phys. Rev., 47: 3452, 1993.

Chapter 18: The Distance Limitation of Gravity

If the sun where not there particles would be bombarding the earth from all sides, giving little impulses by the rattle, bang, bang of the few that hit. This will not shake the earth in any particular direction, because there are as many coming from one side as from the other, from top as from bottom. However, when the sun is there the particles which are coming from that direction are partly absorbed by the sun, because some of them hit the sun and do not go through. Therefore the number coming from the sun's direction towards the earth is less than the other sides, because they meet an obstacle, the sun. It is easy to see that the farther the sun is away, of all the possible directions in which particles can come, a smaller proportion of the particles are being taken out. The sun will appear smaller – in fact inversely as the square of the distance. Therefore there will be an impulse on the earth towards the sun that varies inversely as the square of the distance. And this is the result of large numbers of very simple operations, just hits, one after the other, from all directions.[92]

Richard Feynman, 1965

Gravity Versus The Matter Force

In order to explain that galaxies form due to gravity and yet the universe is expanding at an accelerated rate, there has to be a mechanism whereby the matter repulsion force is stronger than long-range van der Waals gravity at intergalactic distances, while being weaker over distances that correspond with the size of galaxies and lower. Both forces must vary to the inverse of the square of the distance (**$1/r^2$**), so we would naively expect that whatever the ratio of the forces is in nearby space is the same ratio we should see at any distance. So, the physical evidence is not in keeping with simplest case of force dynamics. As stated previously, this is not a fault that only occurs with the theories under consideration, but is something which must be explained by any theory that attempts to account for both gravity and the accelerating expansion of the universe.

While Feynman was discussing the kinetic version of Fatio's Push Gravity Theory of gravity in the above quote, his basic description is essentially correct. Fatio's Theory does in principle lead to a force that varies with the inverse squared of the distance. As has been previously established, gravitation is a push force due to and transmitted by zepton jitter. This produces a long-range van der Waals Force that propagates from zepton to zepton along essentially straight lines. The long-range van der Waals Force due to interactions between electron-like zeptons and proton-like zeptons decreases at a rate proportional to the inverse of the square of the distance ($1/r^2$).

The matter repulsion force on the other hand propagates through polarization of the zero-point field. The antimatter charges of the zeptons orient toward the matter with the matter charges oriented away. The electron-like and proton-like zeptons align in a way so that the zero-point field remains electrically neutral. In some intermediate region of space, matter charges go against matter charges, causing them to deflect away as we see with the repulsive Faraday field line configuration as illustrated in Figure 18-1. This leads to an increase in the pressure pushing the bodies of matter apart. The repulsive Matter Force also decreases at a rate proportional to the inverse of the square of the distance ($1/r^2$).

Figure 18-1 Two bodies of matter surrounded by matter-oriented zeptons in nearby space, oriented in a repulsive configuration.

On the surface it looks like there is an impasse, as the forces should be proportional. There is a difference, however. One force is due to zepton polarization and the other to zepton jitter, a long-range van der Waals Force. The possibility is there that the propagation of the polarization of the zero-point field is more stable than simple random vibrations transmitted zepton to zepton. It is the random nature and linear propagation of the gravitational pressure force transmission due to zepton jitter that ultimately limits the range of the long-range van der Waals Force.

The Solution

We can consider the gravitational force between two stars roughly the size of our Sun. They have a radius of approximately **7 x 10⁸ meters**. To see what happens at distances on a galactic scale, we can imagine they are **4 x 10²⁰ meters** apart, about the radius of the Milky Way Galaxy. We can remember then that gravity effectively acts on the center of mass, so we can draw a line from the outer edges of one star to the center of the other to look at the path that the gravitational force must travel. This is illustrated in Figure 18-2. We can see that the effective path is that narrow if we imagine Star 1 surrounded by a sphere of stars all at the same distance. Each star's gravitational influence can be computed based on the cone the same size.

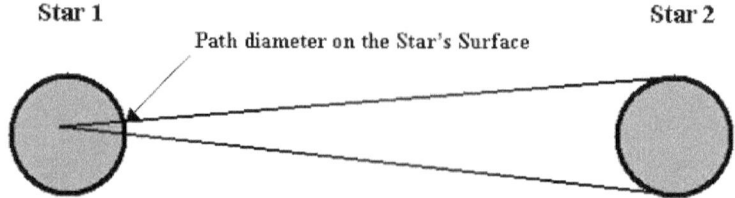

Figure 18-2 The path of the gravitational force from Star 2 to Star 1.

What we then see is that this path gets very narrow as it reaches the distant star. Comparing Figure 18-1 with 18-2, it is obvious that gravity has to propagate over a much narrower region of space. This introduces a certain degree of fragility to the gravitational force propagation, opening the door of possibility a crack. All we need is to identify a reasonable mechanism behind the disruption of transmission of the pressure force.

If we imagine that there are no other bodies of matter within a similar range of distance, we can see that everywhere else on Star 1 will be pushed on equally by the vacuum. That little circle on Star 1's surface ascribed by a cone to the far away Star 2 is the effective area where the gravitational interaction starts. Based on the measurements we started with, the circle at the surface of the first star is about **2.5 millimeters** in diameter. This means that the gravitational force must be conducted along a very long and narrow path. It also means that on a star with a total surface area of over **6 x 10^{18} square meters**, the distant star's gravity affects an area of only about **20 square millimeters**. That is only **3.3x10^{-24}** of the star's surface area. The rest of the **6 x 10^{18} square meters** is being pushed uniformly by other parts of the zero-point field.

The **2.5 millimeter** diameter area equates to a width of **3500** visible red light photon wavelengths, **25 million** hydrogen atoms, or **4 x 10^9** electron-positron pairs across at the pair production energy. If the same stars were instead **1 x 10^{26}** meters apart, on the order of the

182

observable universe, the solid angle at the surface of the star would equate to **100** hydrogen atom diameters across, less than a single visible red light photon wavelength. In order for the gravitational pressure force to be transmitted all that distance, there must be a very high percentage of small high-energy zeptons along the entire path. Given the random nature of the zeptons, that cannot happen.

The first source of gravity degradation we can think of is due to the zeptons themselves. The distribution of zeptons in space is random, so a very short wavelength zepton can end up briefly adjacent to a very long wavelength zepton. In this case, the force will not be transmitted as strongly during a brief time period that that situation exits. The occasional long wavelength zepton, longer than the gravitational path is wide, such as with the red light wavelength in the above example, can cause the long-range van der Waals Force to be completely disrupted, if only briefly. Along the edges of the path, longer wavelength zeptons may be partly out of the path and partly in it, leading to degradation of the force transmission along the edges.

Any electric or matter charge in space can change the local van der Waals Forces slightly, so any dust, gas, plasma, planet, or star in the way will disrupt the gravitational force. The largest effect is likely due to plasma. As we know from studying nebula, they are not made of well-behaved electrically neutral clouds of gas and dust; the constituent particles are ionized, forming plasma. Plasmas produce very large electric and magnetic fields. These fields are many orders of magnitude more powerful than the van der Waals Forces and inhibit their transmission. In a region of space with a large amount of gas, dust, and plasma, the range of the gravitational force will be seriously limited.

If we assume a typical density of interstellar medium of one atom per cubic centimeter, there would be **4 x 10²²** atoms along a centimeter square path between our two stars, which is only one order of magnitude shy of

Avogadro's number. The amount of material over this distance taken all together is sufficient to limit the effective transmission distance of the gravitational force. Over a long distance these various disruptions add up and the force of gravity falls below the matter repulsion force in intensity.

The distance limit of gravity is, however, not due to the loss of gravitational push from the direction of our distant star, but rather a loss of gravitational push from all directions. After all, gravity in the push force model is due to the vacuum of space pushing on a star from the opposite direction of the star interacting with it gravitationally. Gravity stops working when there is no longer a detectable pressure imbalance due to bodies that are large distances away in all directions.

The Range of Gravity
The next question is, what exactly is the range of the long-range van der Waals gravitational force? We know that for spiral galaxies to form, we need an additional force, since there is not enough mass to account for gravity alone to keep a galaxy that size together. Perhaps then, the Milky Way Galaxy was not the best-sized example to start with. In order to eliminate the mattermagnetic force due to angular momentum, we need to consider distance scales the size of nearly spherical elliptical galaxies, with little to no rotation. The galaxy should also not have a well-defined core, as large gravitational structures such as galactic cores have a much longer effective gravitational range. That leaves dwarf elliptical galaxies as the ones that should give us the effective range of gravity due to stars on the size scale of our sun.

Dwarf elliptical galaxies range in size from **500 light years (5×10^{18} meters)** to **10,000 light years (1×10^{20} meters)**. The largest of the dwarf galaxies, **10^{20} meters**, is the most likely to fit with the distance of gravitation limits for typically sized stars. Going back to our model in Figure 18-2, that equates to a gravitational spot size on the surface of the sun of **10 millimeters** diameter.

This is equivalent to **100 million** (10^8) hydrogen atoms across over a distance 10^{30} hydrogen atoms long. Even at that width, there is a lot of opportunity for the force transmission to be diminished or fail.

The result is that the gravitational force between individual stars diminishes at distances that correlate to the size of dwarf elliptical galaxies. The fundamental limit appears to be on the scale of ~10^{20} **meters**. This is not to say that gravity has no effect at great distances. Large collections of objects, such as galactic cores or entire galaxies, have a much larger gravitational footprint than the size of the individual stars comprising them would suggest. This is similar to how the same that clouds of gas and dust act like much larger bodies with respect to transmission of the long-range van der Waals gravitational force than would be expected based on the size of the constituent particles alone.

Conclusion
The matter repulsion force and the long-range van der Waals gravitational force have different mechanisms of propagation. The matter repulsion force is transmitted due to polarization of the zero-point field. Polarization propagates over large distances essentially unaffected by distance. The gravitational push force on the other hand relies on propagation along a line of sight, which becomes vary narrow when two stars are separated at distances the size of elliptical galaxies. Local electric and matter fields and electric charges and bodies of matter throughout space limit the transmission of the long-range van der Waals Force.

This effectively leads to what we observe: that gravity dominates the matter repulsion force on scales the size of medium-sized galaxies and the matter repulsion force dominates at intergalactic distance scales and greater. In this way we can have stars, solar systems, and galaxies formed with the help of gravity, while at the same time having the universe as a whole undergoing an accelerating expansion.

And so, another key point is added to our list.

45) The distance limit of gravity is due to disruptions in the London-van der Waals gravitation force transmission, while the matter repulsion force is not distance limited

[92] R.P. Feynman, The Character of Physical Law, The 1964 Messenger Lectures, pp. 37-39, 1965.

Chapter 19: The Speed of the Zero-Point Universe

If the Sun attracts Jupiter towards its present position S, and Jupiter attracts the Sun towards its present position J, the two forces are in the same line and balance. But if the Sun attracts Jupiter toward its previous position S', and Jupiter attracts the Sun towards its previous position J', when the force of attraction started out to cross the gulf, then the two forces give a couple. This couple will tend to increase the angular momentum of the system, and, acting cumulatively, will soon cause an appreciable change of period, disagreeing with observations if the speed is at all comparable with that of light.[93]

Sir Arthur Stanley Eddington, 1920

History of the Speed of Gravity

The speed of gravity is an important distinguishing feature between Newtonian Gravitational Theory and General Relativity; thus, speed of propagation is one of the important tests of any theory of gravity. Within the scope of Newtonian Gravitational Theory, the speed of propagation of gravity is assumed to be infinite or nearly so, and in General Relativity it is assumed to be the speed of light. For Newtonian gravity to work, there has to be no delay in transmission of the force.

If it looks like the force is coming from point **A** the other body really has to be at point **A**. If gravity is not instantaneous, the force looks like it is coming from point **A** when the body is really at point **B** some distance away. This causes the body at point **B** have a force **F** on it that it not directed toward the Star or whatever body it is orbiting. As the quote from Sir Eddington states, if the speed of propagation of gravity were too slow, orbits would degrade by more than we have observed. The speed of light as it turns out is too slow by many orders of magnitude.

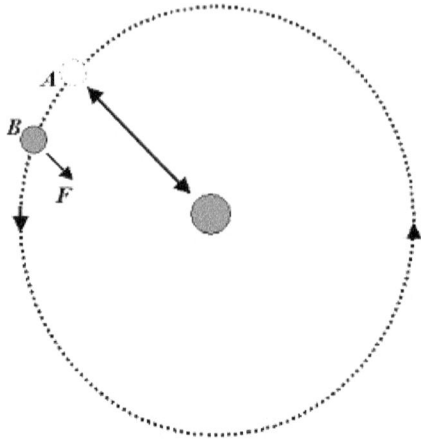

Figure 19-1 If gravity interacts between a star and a planet at point A, but it is really at point B then the force will be in the wrong direction, not directed toward the star.

With General Relativity on the other hand, the speed of gravity is assumed to be the speed of light. It is not initially an issue of General Relativity, but rather of Special Relativity. Around the turn of the last century, a group of relativists, most notably Larmor, Lorentz, Poincarè, and Einstein, were developing electromagnetic equations taking into account the speed of light limitations of photon travel and the motion of bodies. The theory of Special Relativity they developed requires that the speed of light be the same for all observers. Then when Hilbert and Einstein went on to develop the theory of General Relativity, it was taken for granted that the speed of light could not be exceeded. The way they got around the difficulty pointed out by Sir Eddington will be addressed in a moment.

Why this fixation on the speed of light limitation? That comes about for two reasons. One is that if light propagates without a medium, its velocity must be a fundamental property of light. The second problem is causality. If information could be sent faster than the speed of light, it is possible for causality to be violated. In theory, the news of something happening could be transmitted to multiple observers with at least one moving at a high speed, faster than the speed of light relative to the first, and the news can get back to the place the event happened before it happens. The consensus is that if information can be communicated at faster-than-light speeds, then that invalidates Special Relativity.

The problem relates to the question of, what affect the medium of the vacuum has on gravitational force transmission. Does it slow transmission speeds to the speed of light, or can they be faster? Is there a mechanism for transmission or can it be transferred by magic?

> *What is fundamentally new in the æther of the General Theory of Relativity as opposed to the æther of Lorentz consists in this, that the state of the former is at every place determined by connections with the Matter and the state of the æther in neighbouring places, which are amenable to law in the form of differential equations; whereas the state of the Lorentzian æther in the absence of electromagnetic fields is conditioned by nothing outside itself, and is everywhere the same.*[94]
>
> Albert Einstein, 1920

Of course, magic is not a viable physical hypothesis, so we can dispense with that notion. As discussed in the last chapter, there is a medium, the zero-point field, and the gravitational pressure force is conducted through the field point-to-point, zepton-to-zepton as part of a long-range van der Waals Force interaction. The propagation of a force through the vacuum of space must be responsible for some small transmission delay that prevents the speed of gravity from being infinite, but how much is the question.

In Einstein's lecture on æther and relativity, he gives us insight into one of the requirements of General Relativity. Space, the æther, has to somehow know where all the matter is. There has to be some connections between all of the matter in the universe and every place in the æther of space transmitting the information.

Here is the kicker. In General Relativity Theory, in order to prevent rapid degradation of orbits, this information has to be transferred nearly instantaneously, many times faster than the speed of light. This part of General Relativity, the requirement that space has to somehow know where all the matter is, violates the speed of light limitation that is so critical to Special Relativity. Once again we are left with this information having to be transported by some form of magic, nearly instantaneous and without a conduction medium or method of interaction, if we stick with Einstein's theory.

To avoid this problem, supporters of General Relativity theory claim that matter's influence on space is a geometric relation. In other words there is no causal relation - it just is. The metric of space, we are told, is bent by the existence of matter, and bodies of matter follow its curves. If you ask how that information is communicated from matter to space, the quick answer will be that there is no communication. Ah, if we would only believe in magic.

Another point is that Einstein states that the state of the æther of General Relativity is determined by the matter, while the state of Lorenzian æther is free of prior conditioning. The Lorentzian æther model is in this way more ideal as a principle of science philosophy, as theoretically a space, which is free of prior structure, is preferred over one with prior structure. If space knows where matter is in advance, then General Relativity violates one of our important rules: that space should not have any prior structure imposed on it. The bottom line is that General Relativity theory violates the speed of light limitation, even when most of its proponents deny it with their magical excuses.

> *Seeing that gravitation is ether pressure, it does not seem probable that its velocity can be infinite. However that may be, the ability of the ether to transmit pressure and various disturbances evidently depends upon properties so different from those that enable matter to transmit disturbances, that they deserve to be called by different names. To speak of the elasticity of the ether may serve to express the fact that energy may be transmitted at a finite rate in it; but it can only mislead one's thinking if he imagines the process to be similar to energy transmission in a mass of matter.*[95]
> Amos Emerson Dolbear, 1897

The Speed of Long-Range Van der Waals Gravity

Now, how does that tie back to our zero-point universe model? Within the scope of the long-range van der Waals Force model of gravity, the speed of propagation must be faster than light in order to prevent orbits from degrading. Not only that, but since the Newtonian force is a composition of three forces - gravity, matter repulsion, and the mattermagnetic forces - all three fields must propagate faster than the speed of light. That pretty much leaves us with a dilemma. Either fields propagate by magic, or the idea that the speed of light limits speed of propagation is incorrect. Since a physicist should not consider magical mechanisms, that leaves us with one choice.

To get a better analysis of the speed of various interactions through the zero-point field, we should take a step back from the issue of gravity and start over with something simpler, a single photon. Going back to the basic description of a photon discussed in Chapter 5, a photon consists of a sequence of central zeptons associated with a wavefront of associated zeptons in the space around it, as shown in Figure 19-2.

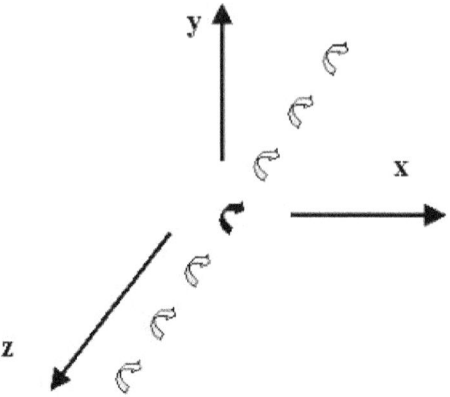

Figure 19-2 A photon propagating in the x direction as indicated by the black central zepton, which is surrounded by zeptons oriented and rotating with respect to the photon's electric and magnetic fields.

As a central zepton rotates, it produces electric and magnetic fields that propagate outward at a 90-degree angle to each other, while also perpendicular to the direction that the photon is moving. That means that there are zeptons in the space around the central zepton that rotate and orient themselves with respect to the central zepton. The field radiates outward from the central zepton and declines in strength at a rate proportional to the inverse of the distance squared ($1/r^2$).

We know that in a given frame of reference, the time t it takes for a photon to travel from point **A** to point **B** in that frame is determined by the distance between those points and the speed of light. From the above model, we can see why the photon travels at the speed of light in a vacuum, as it goes back to the idea that the zeptons are more fundamental than the photon. The time that a zepton can exist with regards to Planck and Heisenberg limits equates to a wavelength equal to one half the photon wavelength for that energy. The speed of light is not a fundamental property of light; it is a fundamental property of vacuum fluctuations. It is the maximum extent of travel for a zepton during its brief existence, its wavelength. You may recall the following relations from Chapter 2 shown in Equation 19-1, relating the time **Δt** with frequency **ν**.

Equation 19-1

$$E = \frac{\hbar\omega}{2} = \frac{h\nu}{2} \equiv \Delta E \Delta t = \frac{h}{2}$$

Then we can recall from Chapter 4, the relationship between frequency of the zepton and its wavelength, with the zepton wavelength being half the photon wavelength as shown in Equation 19-2.

Equation 19-2

$$\frac{h\nu_z}{2} = \frac{hc}{2\lambda_z} = \frac{hc}{\lambda_{ph}}$$

We also know that the energy of the photon originates at point **A** and is absorbed at point **B**. A photon never loses a part of its energy; it is always absorbed all at once. The magnetic and electric fields do not exist prior to the origination of the photon, nor after the photon is absorbed. The fields are also extinguished each half wavelength along with the central zepton. Even so, the electric and magnetic fields are detectable at more than half a wavelength distance off to the side of the photon's path. But remember that the half wavelength distance is related to the speed of light. That means that the electric and magnetic fields of the photon are propagating faster than the speed of light, or they would not be detectable and would not interact with other, more permanent fields.

We can see this more precisely by considering a double slit experiment with a single photon as illustrated in Figure 19-3. If we have a photon traveling between point **A** and point **B** with nothing in the way, it will reach point **B** in a time **t** that is exactly in keeping with the speed of light. Now we introduce a wall between point **A** and **B** with two slits some distance apart, equidistant from the straight-line path. If we then send a photon through the slits from point **A** to point **B,** it will still take the same time **t** even though the photon has to travel through slits offset from the direct path.

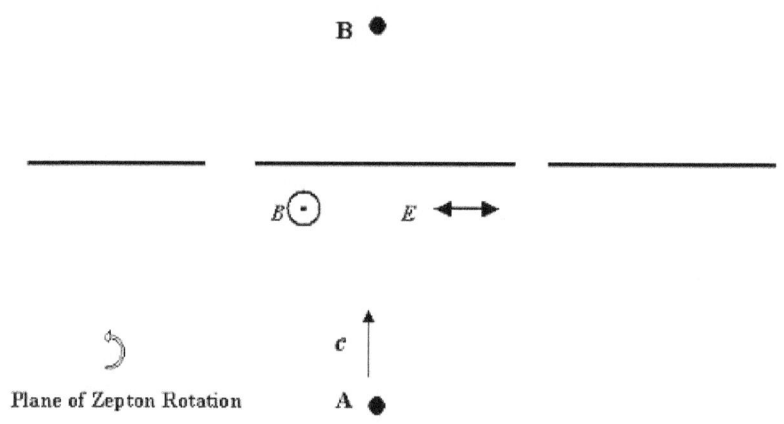

Figure 19-3 A photon moving from point A to B polarized in the plane of the surface of the page.

The information of the photon, the magnetic and electric fields, travels through the slits, but the propagation of those fields in a direction perpendicular to the direction of the photon happens at a rate much faster than the speed of light. If the propagation were not much faster than the speed of light, the duel split experiment would not work in the first place, as the fields would only go out a half wavelength to either side of the photon's path. Even more so, light would not show diffraction patterns, as those patterns exist entirely at distances greater than half a wavelength from the photon. Note that the matter and mattermagnetic fields also propagate much faster than the speed of light in precisely the same manner.

The zeptons that make up the photon in this case are shown rotating in the plane of the page. This produces an electric field **E** that oscillates left and right and a magnetic field **B** that oscillates in and out of the page. If you imagine that the wavelength is equivalent to the diameter of the dots at points **A** and **B**, then if the propagation of the electric and magnetic fields is equal to the speed of light, they can only propagate out a distance equivalent to the radius of the dots. The fields, and thus the photon, could never make it through the slits if the speed of field propagation was the speed of light.

Another way to look at it is related to a zepton's rotation. If a zepton has to rotate 180 degrees during its life, then the velocity of propagation in the direction of rotation is the speed of light. This is also true when matter moves through space and an inertial mattermagnetic field develops around it. As the velocity of matter approaches the speed of light, the rotating zeptons of the surrounding mattermagnetic field approach 180-degree rotation during their existence. They cannot rotate farther or faster than that. The zepton rotation concept is illustrated in Figure 19-4. The maximum distance **d** a zepton can occupy in space during the time it exists is equal to its wavelength **λ**, which is equal to the speed of light **c** divided by the zepton frequency **v**.

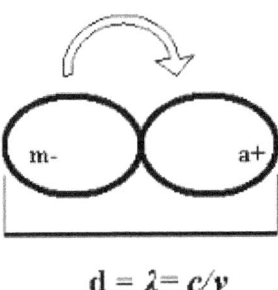

$$d = \lambda = c/v$$

Figure 19-4 A zepton rotating 180 degrees occupies a maximum distance **d** equal to its wavelength.

The Speed of Fields

When it comes to propagation of a linear electric or matter field, we have imagined that zeptons form chains of dipoles, which are reminiscent of field lines, polarizing the field of the vacuum. In a more realistic scenario, however, zeptons only have to rotate a fraction of a degree during a fraction of the time they exist in order to produce the field. Keep in mind that the mass-energy density ratio between the zero-point field and permanent matter is on the order of 10^{95}. The ratio of charges in the zero-point field to permanent charges is similarly high, perhaps not to that degree, but for all practical purposes, nearly infinite. It takes a fairly insignificant change in average zepton orientation to produce a huge electric or matter field. For example, if the average polarization only requires $180/10^{20}$ degrees of rotation then the polarization can propagate at 10^{20} times the speed of light. Note that this would require that the vacuum have a minimum of 10^{20} times the charge of the permanent charges involved in the interaction.

Magnetic field propagation is similar as it propagates along the axis of rotation. Zeptons only have to rotate a fraction of a degree in order to produce a huge magnetic field, and this can occur very quickly, much faster than the speed of light. As above, if the average rotation producing the magnetic field only requires $180/10^{20}$ degrees of rotation, then the magnetic field can propagate at 10^{20} times the speed of light. In the zero-point field, both linear polarization and rotation can easily propagate through space at speeds far greater than the speed of light. The $10^{20}X$ example above may be nowhere close to the real propagation speed, which has not been determined.

Since gravity is a long-range van der Waals Force, it is the result of electrostatic and matterstatic repulsion. The zeptons are not moving in response to an external field, but move more or less randomly in response to neighboring charges. As with the basic field interactions, the long-range van der Waals Force can be transmitted without requiring the zeptons to rotate the complete 180 degrees. This means that, as with the basic field forces, the rate of propagation of the gravitational long-range van der Waals Force is many orders of magnitude greater than the speed of light.

There have been previous estimates of the speed of gravity based on the speed needed to prevent orbits from decaying any more than we have measured. Laplace made the first serious estimate in 1825.[96] The value he derived, based on non-relativistic mechanics, was that the speed of gravity had to be at least **7 x 10⁶** times the speed of light. This minimum multiplier has continued to increase as we acquire increasingly precise data. In his 1998 paper, Tom Van Flandern states that the speed of gravity is at a minimum **2 x 10¹⁰** times the speed of light.[97] This paper by Van Flandern is one of the best at explaining the speed of gravity problem, anyone who wants more detailed information on this topic should review it.

Conclusion

When proponents of the theory of General Relativity claim that the speed of gravity is the same as the speed of light, they are perpetuating a myth. Whatever link they presume there is between matter and space, informing space where the matter is located, that information must be passed along at a speed much greater than the speed of light, or the orbits would have degraded long ago. Every theory of gravity requires that gravity propagates many orders of magnitude faster than the speed of light.

Merely the existence of light diffraction and the results of the double slit experiments tell us that the electric and electromagnetic fields must also propagate at a rate much faster than the speed of light. Likewise so must the matterstatic and mattermagnetic forces. The rate of propagation of all the electro-matter forces can be readily explained by considering the amount of rotation a vacuum fluctuation has to go though for those fields to propagate. Only a small amount of rotation is necessary to propagate those basic fields, while at the same time, the motion of permanent matter and photons is still limited by the speed of light.

The speed of propagation of fields including electro-static, matterstatic, electromagnetic, mattermagnetic, and gravity is at least **2 x 10^{10}** times the speed of light based on astronomical data. It could easily be many orders of magnitude greater than that, since a more precise estimate for the maximum rate of propagation has yet to be determined. That number is ultimately dependent on the ratio of permanent charges to virtual charge dipoles in the vacuum.

Keeping up with our tally of findings:

46) The speed of gravity is much greater than the speed of light
47) The speed the electro-matter field propagation is much greater than the speed of light

[93] A.E., Eddington, Space, Time and Gravitation, original printed in 1920, reprinted by Cambridge Univ. Press, Cambridge p. 94, 1987.

[94] A. Einstein, "Æther and the Theory of Relativity" Address delivered on May 5th, 1920, at the University of Leyden, Germany.

[95] A.E. Dolbear, Modes of motion; or, Mechanical conceptions of physical phenomena Boston, Lee and Shepard , 1897.

[96] P. Laplace, Mechanique Celeste, volumes published from 1799-1825, English translation reprinted by Chelsea Publ., New York pp. 642-645, 1966.

[97] T. Van Flandern, The Speed of Gravity What the Experiments Say, Phys Let A, 250:1-11, 1998.

Chapter 20: The Origin of the Speed of Light

The physics of the future, of course, cannot have the three quantities ħ, e and c all as fundamental quantities. Only two of them can be fundamental and the third must be derived from those two. It is almost certain that c will be one of the two fundamental ones.[98]

Paul Dirac, 1963

London-Van der Waals Torque

If the speed of light is determined by the time it takes a zepton to rotate, what causes the zepton rotation to be this particular value? To understand we first need to consider a zepton rotating in the vacuum. When it rotates, it causes other nearby zeptons to rotate. This induced rotation, however, does not come without a cost. It takes energy to induce rotation in the other zepton dipoles. This is energy we generally do not factor into our equations, as it gets ignored on both sides of the equations, but it is there nonetheless.

The zero-point field's resistance to rotation produces a drag on the rotating central zepton of the photon preventing it from rotating faster. Conversely, if there were no zero-point field, a single zepton would rotate infinitely fast. The zero-point field induces a torque on rotating zeptons slowing them down, ultimately to a degree that equates to the speed of light. The dipole-to-dipole interactions producing this torque are fundamental to London-van der Waals Forces, so it is more properly referred to as a London-van der Waals torque.

In a vacuum in free space, the distribution of zeptons is uniform and constant, which produces a constant torque, and hence a constant speed of light. The speed of light is not a fundamental constant; the torque due to the zero-point vacuum fluctuations is what is fundamental and constant in a vacuum in free space.

The speed of light limit comes about when matter or energy is moving through space. Matter is accompanied by rotating zeptons that form an inertial field that is rotating, with the rate of rotation limited by the London-van der Waals torque of the vacuum. A photon is entirely composed of rotating zeptons, thus its speed is limited to the speed of light as the London-van der Waals torque slows its rotation.

This takes us back to another old problem. What happens when the speed of light slows down in a transparent medium, such as glass or water? It has long been troublesome to physicists to think of the value of c changing just because light is traveling through a body of matter, particularly if the speed of light is a property of light and not the medium of travel. If c changes, scientists would be at a loss as to what could cause it. In a typical glass, for example, the speed of light may only be 66% of its usual value. Most physicists ignore the fact that within the Standard Model there is no reasonable physical model that explains what is going on. Oh, there is an explanation. The photon energy is said to be absorbed and then re-emitted. You could call it the catch-and-release model. Why would some photons not go through uncaught, as gamma rays do through lead? Why would they all get caught the same amount of the time? Why would they get released at all? In real-world physics, once a photon is absorbed, it is absorbed completely. Some physicist was really grasping when he came up with this one. The catch-and-release model is not in keeping with standard physics of photon interactions. It is utter nonsense.

According to the London-van der Waals torque model, the amount of torque increases in the presence of matter since the vacuum fluctuations respond to both the matter and the photon, making for a greater amount of drag on the photon's central zepton rotation. The London-van der Waals torque also directly converts to a real measurable torque as seen in the refraction, or bending, of the light as it enters glass, water, or other matter medium. The value of the torque changes, which leads to the derived constant **c** changing. Dirac guessed wrong; **c** is not fundamental.

Conclusion

We have also seen that the speed of light is not, in and of itself, a fundamental quantity. It is rather the result of a torque on a rotating zepton due to the zero-point field's London-van der Waals torque. Given that information, we should now be able to fundamentally derive the speed of light from first principles. This can be accomplished with a strictly classical type zero-point field model of the universe.

In this chapter the new key points were:

48) The speed of light is derived from the London-van der Waals torque of the zero-point field
49) Changes in the speed of light in a medium are due to changes in the London-van der Waals torque

[98] P.A.M. Dirac, *The Evolution of a Physicist's Picture of Nature*, Scientific American 208:5 pp 45-53 (May 1963).

Chapter 21: Æther and Michelson-Morley

> *That a body may move in the ether for an indefinite time without losing its velocity has been stated as reason to believe the ether to be frictionless.*[99]
>
> Amos Emerson Dolbear, 1897

Old Æther

The denigration of æther theory and anyone who supports it is one of the great travesties of the science of the last century. For as we are learning, the zero-point field is critical to our understanding of everything. Today, students are commonly taught that æther theory is archaic, that it was disproved by the Michelson-Morley experiment, and anyone who espouses such a theory should be looked at with scorn and ridicule. Even mentioning the word æther in a scientific paper is often *ad hoc* grounds to reject the article. Consequently, æther theories are disguised with terms like; *quantum mechanical oscillator, vacuum fluctuation, London-van der Waals Force of the vacuum, Cosmological Constant*, and yes of course, *zero-point energy*. As to the last, it is getting to where the term *zero-point energy* is treated about the same as æther, as well it logically should. After all, they are equivalent, as Nernst figured out long ago.[100]

> *To deny the æther is ultimately to assume that empty space has no physical qualities whatever. The fundamental facts of mechanics do not harmonize with this view.*[101]
>
> Albert Einstein, 1920

As many scientists did before him, even Einstein recognized that if one is to find a mechanical explanation of force interactions, there must be æther, at least he admitted such in his 1920 presentation.[101] Beyond that, his position must have been a very complex one, since if æther is zero-point energy, then General Relativity cannot be due to all the energy in space. The zero-point field theory must be ignored. The Einsteinian æther must also communicate with all matter at the same time faster than the speed of light while bending space-time - quite a feat. This inconsistency is at the heart of why his efforts to unify the forces failed, as well as the efforts of others who followed in his footsteps. In the absence of a mechanical force, one is left with only magic, and magic is unacceptable as a scientific theory. As we will see, his failure to incorporate vacuum fluctuations into his theories of relativity led to many other errors, most notably those theories as a whole. It is not at all clear what Einstein was trying to say when he used the term "æther" in 1920. Nonetheless, his opinions on æther are often quite thought provoking and quote worthy.

Æther theory underwent a surge in popularity in the mid-1800s, once Faraday discovered the electromagnetic properties of light, and physicists decided that light needed a medium to propagate through. Little did they know that not only was their line of reasoning correct, but photons are entirely composed of the æther. Unfortunately, æther incurred a huge setback due to the Michelson-Morley experiment.[102] There is a lot of erroneous information about the experiment. The way the history of the experiment is taught is at best disingenuous and at worst an act of sabotage on physics in general.

Maxwell's Conjecture
At the time, it was thought that light propagated through space much in the way sound does. So it was expected that light would have a preferred direction relative to the æther. An æther tailwind, for example, would allow light to move faster than with a headwind, and a photon crossing an æther stream would take a different amount of time than one going back and forth with the stream. It was expected that photons moving across the stream would take longer. The photons going across the stream were thought to have to travel the distance equivalent to the hypotenuse of a right triangle where one side is the distance across the stream at a

fixed point in relation to the æther, and the other side is the distance traveled by the Earth relative to the æther during the time it takes light to travel back and forth. James Clerk Maxwell conjectured in 1879 that this experiment might allow us to calculate the speed of the Earth through the æther.[103] The basic geometry is illustrated in Figure 21-1.

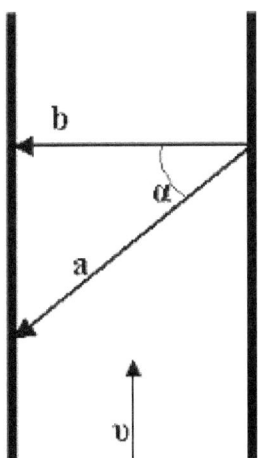

Figure 21-1 A beam of light crosses between two walls along path **b**, but because of the æther direction and velocity **u** shown by arrow, the effective distance is thought to be **a**. And angle **a** is shown between the two paths.

You can get the idea by imagining yourself swimming across a stream. You have to swim at an angle into the current to make it across in a straight line. This angle would be like **a** in the illustration. You would try to swim along line **a**, but your actual path is **b**. The time it would take for you to swim across the stream would be the same as if you swam distance **a** in a pool with no current. If the angle **a** were 45 degrees, the length of **a** would be about ~1.4 times **b**. Your velocity swimming across the stream would then be your normal speed swimming in a straight line without a current divided by ~1.4. So, the current in the stream makes you appear to swim slower. Many physicists thought the same thing would happen to light.

If you then swim the same distance as **b** against the current, it will take you a little longer than if there were no current. But coming back, it will take a little less time. Those two time differences cancel out, so that your average speed is your normal average speed when there is no current present. This is basically how the Michelson-Morley experiment was set up.

The Michelson-Morley experiment assumes that if the speed of light is the same in both directions, then there is no æther. It was thought that since Earth is obviously moving through space that the speed of light will be different depending on whether it is going across the æther or with it.

What the experiment showed was that the speed of light is the same in both directions, the constant **c**. It was thought that this meant that there was no æther. This experimental result, however, does not say anything about the existence of æther, only that their theory of light propagation being somehow affected by a velocity relative to the æther rest frame was incorrect. It shows conclusively that the speed and wavelength of light are both constant when seen by an observer in the same frame of reference as the light source, in other words not moving relative to the speed of the light source.

Lorentzian Æther

George Fitzgerald is credited as the first to propose in 1889 that the problematic result of the Michelson-Morley experiment could be accounted for if length were contracted in the direction relative to the motion of the æther.[104] Hendrik Lorentz independently proposed the same idea in 1892.[105] It was Joseph Larmor in 1897 and 1900 who first published the modern form of what would come to be known as the Lorentz equations, which included the time dilation term.[106,107] Those equations had first been published by Woldemar Voigt in 1887, but it was in a somewhat different context.[108]

Lorentz published his version of the Lorentz equations in 1899 and 1904, initially being unaware of Larmor's work.[109,110] It was Poincaré who dubbed them to be Lorentz's equations, and the name stuck.

The Lorentzian æther theory relied on length contraction and time dilation to explain the lack of a difference in light speed with respect to its direction relative to the æther. On a broader scale, it became a new way of transforming distance and time intervals between different frames of reference, where one frame is moving relative to the other. In his theory, the photons crossing the stream reached the detector at the same time as the other beam did, because the length was shortened. This was the invention of the Lorentzian æther discussed earlier. This theoretical æther had properties that allowed it to match the experimental results as interpreted by the physicists of the time.

> *The principle of relativity, according to which the laws of physical phenomena must be the same for a stationary observer as for one carried along in a uniform motion of translation, so that we have no means, and can have none, of determining whether or not we are being carried along in such a motion.*[111]
>
> Henri Poincaré, 1904

Poincaré was the one responsible for the term Principle of Relativity as it was a term he used when teaching at the Sorbonne. He published a variation of Lorentz's work in 1904, further developing it into a form nearly identical to Special Relativity, except that he maintained the idea that motion was relative to an æther rest frame.[111] A year later, Einstein came up with Special Relativity by doing away with the æther rest frame and requiring that the speed of light be constant relative to all observers.[112] Both their theories retained the same length contraction and time dilation as Larmor and Lorentz. The differences in Lorentzian Relativity and Special Relativity can be quite subtle, but nonetheless important, as it is always necessary to have a standard reference frame when computing time differences. Without a standard reference frame, one has no way to know which observer will have his or her time rate changed.

The Transformation Math

To better understand the details it is best to at least go through a simple mathematical introduction of the concepts. Under the old Newtonian/Galilean view, time is expected to be measured the same in a moving frame of reference as we would see in our own stationary frame of reference. So if one frame is moving in the *x* direction along the *x* axis, we only need to adjust our measurements on that axis with the following Equation 21-1, where the prime marks indicate the moving frame, *v'* the velocity of the moving frame and *t'* the time in the moving frame, which is the same as the time *t* in our frame.

Equation 21-1

$$x = x' + v't'$$

This is really easy to understand, since if we start with two frames of reference aligned such that $x = x' = 0$, and the second frame is moving at **10 meters per second,** after **2 seconds** it will have moved **20 meters**. So after the movement, where *x* is at zero in the fixed frame, *x'* will be **20 meters** in the moving frame.

If we return the Lorentz solution of the Michelson-Morley experiment, it was thought that the distance had to be shortened in the moving frame of reference. This requires a correction factor when we consider velocities at or approaching the speed of light. The Lorentz transformation for a frame moving along the x-axis is as follows in Equation 21-2

Equation 21-2

$$x = \frac{x' + vt'}{\sqrt{1 - v^2/c^2}}$$

It is easy to see from the equation that the v^2/c^2 term goes to **1** as the relative velocity approaches the speed of light. The denominator then approaches **0**, as it is the square root of (**1** - ~**1**). In this way the distance in the moving frame (**x'**+ **vt**) appears much smaller than the distance as viewed from the stationary frame. The length in the moving frame is said to be contracted, and hence the term Lorentz contraction. The term in the denominator is often abbreviated using the symbol **γ** as shown in Equation 21-3, simply as a matter of convenience when writing relativistic equations.

Equation 21-3

$$\gamma = \frac{1}{\sqrt{1 - v^2/c^2}}$$

The other key part of the transformation relates to time dilation. Time, or more accurately clocks run at a different rate when they are moving relative to an observer. They move slower. It is important to note, however, that the clocks in the Michelson-Morley experiment were not moving relative to the Earth's surface, so there was no time dilation in that experiment. The Lorentz transformation equation for time is shown in Equation 21-4.

Equation 21-4

$$t = \frac{t' + vx'/c^2}{\sqrt{1 - v^2/c^2}}$$

Once Again, It is Not a Kinetic Interaction

Let us think a moment about this. The problem was thought to be about the transmission of light through æther, but the answer is relativistic reference frame transformations, Special Relativity. At some point, the problem switched from being a transmission of light through æther problem to a frame transformation problem. Those are two entirely separate problems, which should be treated separately. The way that these two problems have been lumped together is one principle source of much of the misunderstanding of æther. What we ultimately found out is that the Michelson-Morley experiment told us that scientists of the day were confused about how to compute frame transformations involving light speed when attempting to only use Newtonian/Galilean transformation mechanics.

It is easy to see why Michelson's conjecture about æther failed, if we consider the fundament nature of the photon. When a photon is first emitted from a source, there is a zepton that rotates 180 degrees, producing an electric and magnetic field. Other zeptons rotate and orient with respect to the central pair forming the wavefront. Successive central zeptons follow as the photon propagates through space, until at some point it is absorbed. If the photon is emitted into a vacuum, the wavelength times the frequency will equal the speed of light. The light also has the proper energy and spectral characteristics of the light source as normally seen from that type of source in its rest frame. In order for Michelson's version of light transmission through the æther to be correct, other vacuum fluctuations would have to push on and move the photon's central zepton in the direction of motion of the æther.

This classical misperception of æther came about because physicists where attempting to apply kinetic gas theories to zero-point vacuum fluctuations. Vacuum fluctuations do not interact kinetically as we already discovered while considering Fatio's theory and the Casimir Effect. The individual zeptons that make up the photon do not exist long enough to interact kinetically, being constrained by the Uncertainty Principle. Their interactions are due to charge dipole orientation and rotation. Æther as a whole is charge neutral, and the vast majority of interactions are purely random. There is no mechanism for æther to push a photon to one side as envisioned through kinetic gas theory. The only true interaction mechanisms, such as inertia and the London-van der Waals torque, are already accounted for in the physical model. To what extent the æther does push, it pushes equally in all directions and from all sides.

One of the more exasperating things about this whole situation is that Einstein and Hopf in their landmark paper in 1910 showed that "there is no frictional force acting on a dipole or atom moving with constant velocity in the vacuum" where the vacuum in question is populated with zero-point vacuum fluctuations.[113] Just a few short years after publishing his theory of Special Relativity Einstein proved that length contraction and time dilation were not necessary to explain the Michelson-Morley result. In so doing he also proved that the Micheslon-Morley result does not disprove æther, but rather, if the æther is zero-point energy, the Michelson-Morley result is what is expected.

If we examine the null result of the Michelson-Morley experiment there are three possible solutions:

1. There is an æther, it behaves in a manner consistent with kinetic gas theory, and the Lorentz equations - including length contraction and time

dilation of space - are needed to get the correct solution,

2. There is an æther, it is zero-point energy which is consistent with Einstein-Hopf, and the result is as expected, or

3. There is no æther, and the Michelson-Morley result is as expected.

Of course the only proper answer is 2: there is an æther, the zero-point field, and as Einstein himself showed along with Hopf, length contractions are unnecessary to explain the Michelson-Morley result. The Lorentz equations simply do not apply to the Michelson-Morley experiment. As long as the light source and sensor are in the same frame of reference, the light energy and wavelength at the source will be the same as the energy and wavelength at the detector, and the time it takes for the photon to travel a two-way trip between the two will be determined by the speed of light as measured in that frame of reference. This is true of any light source and sensor in a single frame of reference. There is no length contraction or time dilation of space. The angle relative to some perceived æther velocity is simply not relevant.

Maxwell's Conjecture Applied to Inertia

We can approach it from a different direction. If Maxwell's conjecture were correct, then bodies of matter as well as photons must experience the same effect. If you can first imagine a body of matter moving against the æther, it would need to have some additional force behind it to overcome the æther drag. Since there is no measurable force, it would have to be in the form of something like an extra component to the inertial force pushing it along, forgetting for a moment everything we discussed in the chapter on inertia and thinking of it hypothetically. Then, if a body were moving with the æther, it would have an æther tailwind, which would mean it would need less inertia than when the body moves into the æther.

If the body had different inertia when going in different directions, it would take a different amount of force to cause it to change direction when moving in one direction versus the æther than when moving in a second direction. This would show up as an inconsistency in mechanical force law. We can consider a semi-circular particle accelerator, a racetrack configuration. If Maxwell's conjecture were true in the general case, then the strength of the magnetic fields in the cyclotron would have to be different in each curve in order to compensate for the direction of the æther.

This simply does not happen. Maxwell's conjecture is wrong whether you are applying it to bodies of matter or photons. The two-way speed of the light is the same when measured in a given rest frame, and it has nothing to do with length contraction or time dilation. The æther, the zero-point field, the zeptons, simply do not interact with matter in such a way as to produce drag. A change in direction is simply not a change in reference frame. With respect to the Michelson-Morley experiment, relativity theories are an unnecessary explanation to a non-existent problem. Dolbear understood this back in 1897 because the planets show no signs of being subjected to æther drag.[99] Maxwell and his contemporaries should have recognized this as well.

Paradox of Special Relativity

If one takes the stance of mainstream physicists that æther does not exist, there cannot be a universal rest frame. There can be no absolute standard to measure velocity and position against. All measurements then must be made relative to something else, with no universal standard. All measurements must be relative, not absolute. Special Relativity supporters are all nodding their heads in agreement right now.

If that is the case, what is the standard that length contraction is based on? If we measure the length in the laboratory, it will be the standard length as measured in the laboratory's rest frame. If we do the Michelson-Morley experiment and like them assume that the relative velocity between the Earth and Sun, **~30 km/s** is what they were detecting, the amount of length contraction as required by special relativity equates to that velocity. If instead we use the relative velocity between the Sun and the center of the Milky Way Galaxy, ~**225 km/s**, the contraction must equate to that velocity. Then if we look at our speed relative to the Andromeda Galaxy, it is approaching us at **120 km/s**, so space has to expand to account for that, while at the same time contracting from all the galaxies moving away from us.

What special relativity requires is that all of space must be contracted or expanded on a scale from objects moving away at the speed of light, to stationary objects, to objects moving towards us at the speed of light, and everywhere in between all at the same time. How does space know what you are trying to measure at any given instant and how it needs to contract to match the observation? The concept of length contraction in Special Relativity is illogical to the point of being ludicrous. Einstein should have been laughed out of the room when he proposed it.

Conclusion

When these otherwise brilliant scientists were puzzling over the result of the Michelson-Morley experiment it is amazing that none of them asked the question, what if Maxwell's conjecture is simply wrong? Or, what if the æther is incapable of pushing on photons? There was another answer to the problem all along, the simplest possible answer. Instead they went with a different theory, a theory that was far more complicated and, most importantly, a theory that was unnecessary.

Now that we have beaten that horse to death, we can point out the key points of this chapter

   50) The æther is the zero-point field
   51) The zero-point field does not interact kinetically with light or matter
   52) The Lorentz transformations were never necessary to solve anything related to the Michelson-Morley experiment

53) The two-way speed of light is constant when the source and detector are in the same frame of reference

[99] A.E. Dolbear, Modes of motion; or, Mechanical conceptions of physical phenomena Boston, Lee and Shepard, 1897.

[100] W. Nernst, Über einen Versuch, von quantentheoretischen Betrachtungen zur Annahme stetiger Energie Änderungen zurückzukehrenVerh. Dtsch. Phys. Ges., 4, pp. 83-116 (1916). (translations and useful commentary from P.F. Browne, The Cosmological Views of Nernst: an Appraisal, APEIRON Vol. 2 Nr. 3 July 1995).

[101] A. Einstein, "Æther and the Theory of Relativity," Address delivered on May 5th, 1920, at the University of Leyden, Germany

[102] A.A. Michelson, E. W. Morley "On the Relative Motion of the Earth and the Luminiferous Ether". American Journal of Science **34**: 333–345 1887.

[103] J. C. Maxwell, "On a Possible Mode of Detecting a Motion of the Solar System Through the Luminiferous Ether" Nature, 1880, Vol. XXI, pp. 314, 315.

[104] G.F. FitzGerald, "The Ether and the Earth's Atmosphere", Science 13 (328): 390,doi:10.1126/science.ns-13.328.390, 1889.

[105] H.A. Lorentz, "The Relative Motion of the Earth and the Aether", Zittingsverlag Akad. V. Wet. 1: 74–79, 1892.

[106] J. Larmor, "On a Dynamical Theory of the Electric and Luminiferous Medium, Part 3, Relations with material media", Phil. Trans. Roy. Soc. 190: 205–300, 1897. doi:10.1098/rsta.1897.0020.

[107] J. Larmor,, Aether and Matter, Cambridge University Press, 1900.

[108] W. Voigt, "On the Principle of Doppler", Nachrichten von der Königl. Gesellschaft der Wissenschaften und der Georg-Augusts-Universität zu Göttingen (2): 41–51, 1887.

[109] H. Lorentz, "Simplified Theory of Electrical and Optical Phenomena in Moving Systems", Proceedings of the Royal Netherlands Academy of Arts and Sciences 1: 427–442, 1899.

[110] H. Lorentz, "Electromagnetic phenomena in a system moving with any velocity smaller than that of light", Proceedings of the Royal Netherlands Academy of Arts and Sciences 6: 809–831, 1904.

[111] H. Poincaré Bull. Sci. Math., (2) 28, 317-, November 1904. English translation: Bull. Amer. Math. Soc. 37, 2000, 25 - 38.

[112] A. Einstein, "Zur Elektrodynamik bewegter Körper", Annalen der Physik 322 (10): 891–921, 1905.

[113] P.W. Milonni, The Quantum Vacuum, Academic Press LTD, London 1994, p. 60

Chapter 22: Relativity is Not Special

In the first place, we shall ascribe to each electron certain finite dimensions, however small they may be, and we shall fix our attention not only on the exterior field, but also on the interior space, in which there is room for many elements of volume and in which the state of things may vary from one point to another. As to this state, we shall suppose it to be of the same kind as outside points. Indeed, one of the most important of our fundamental assumptions must be that the ether not only occupies all space between molecules, atoms or electrons, but that it pervades all these particles. We shall add the hypothesis that, though the particles may move, the ether always remains at rest. We can reconcile ourselves with this, at first sight, somewhat startling idea, by thinking of the particles of matter as of some local modifications in the state of the ether. These modifications may of course very well travel onward while the volume-elements of the medium in which they exist remain at rest.[114]

Hendrik Lorentz, 1906

There are two guys, Ted and Fred, who are traveling away from each other at a high rate of speed. Under Special Relativity, there is no standard frame of reference, so neither of their positions is special. They both carry identical clocks. Under Special Relativity they are supposed to experience time dilation effects. Because neither of them is special, but they both think they are, Ted thinks Fred's clock will run slow, while Fred thinks Ted's clock will run slow. So which clock runs slow in the clock's frame of reference? The answer is neither, as that is the only way Special Relativity can be correct in both cases. Each clock must run at the same rate in its frame of reference. Second question:

which one, Ted or Fred, actually sees the other one's clock appear to run slow? The only answer is both, since the classical Doppler effect and symmetry requires it. So, if we try to implement Special Relativity with respect to time dilation, we find that it requires symmetric asymmetry, which is impossible. Either the result is asymmetric with one clock being slower than the other, meaning there is a preferred rest frame, or the results are symmetric so there is no clock rate slowing, contradicting Special Relativity theory and experimental evidence. The Special Relativity postulate requiring that there is no preferred frame of reference, or more specifically that the speed of light is the same for each observer, is false.

Light's Frame of Reference
To get at the heart of the whole relativity frame transformation issue, we need to step back a bit, and once again recall how light propagates. Going all the way back to Chapter 5, we see that light, the so-called photons, are actually made entirely of vacuum fluctuations, zepton dipoles. So guess what, the speed of light is constant with respect to the zero-point field in a vacuum. Light's frame of reference is the zero-point field and only the zero-point field. The speed of light is determined relative to the zero-point field.

This concept is not something new. While Maxwell was figuring out that electricity and magnetism worked in a consistent way that could be combined into a single theory, he also included light and optics. He found that the speed of light is constant with respect to the æther.[115] That was the accepted physics model up until Einstein challenged the æther theory and included the postulate that the speed of light was the same as viewed by every observer. There was and still is no evidence to support those *ad hoc* claims. Light is not transmitted based on what is going on with some random observer. Light is made of æther and moves with respect to æther, the zero-point field, always. The zero-point field is light's frame of reference.

The follow-up question is then, what is the zero-point field's frame of reference? If we think about rotating zeptons forming a frame of reference, is it really possible that they can rotate 180 degrees over an infinite range of relative velocities, such that there are an infinite number of possible rest frames that are physically equal to a universal rest frame with respect to light transmission? Or, is there one frame of reference where zeptons are generally at rest where they have the greatest degree of freedom to rotate 180 degrees in any direction? It would be in that frame of reference, and only that frame, where light is transmitted and can be measured at its maximum velocity. The first concept of a multivariable field of zeptons is a physical impossibility. As we discovered with Special Relativity, if space has to simultaneously behave in a certain way for observers in an infinite number of frames of reference, the theory requiring that is not correct. This difficulty tips the balance in favor of the single fixed frame model, and that requires that the speed of light is not constant except in that fixed frame in which it is being transmitted.

As for the question of where is the fixed frame of reference, we can say with a very high degree of certainty that the Earth is not it. The zero-point field reference frame cannot orbit a star or even a galaxy. There is, however, one exceptionally good candidate from among all the things that are known: the rest frame of the cosmic background radiation.

The Space Contraction Paradox
Speaking of the flat space concept, one of the difficulties with frame transformation theory is the length contraction and time dilation hypothesis with respect to space. Length contraction originally was considered with respect to light transmission, and we most certainly can and absolutely must consider differences in photon wavelengths in different rest frames. We can even consider whether or not bodies of matter are contracted when they are moving relative to the zero-point field's rest frame, the universal rest frame. What does not

make sense is that the underlying geometrical framework of space can be contracted. Based on relativity theory, if we consider one rest frame moving relative to another, the unit distances **x** and **x'** are related by the following Equation 22-1.

Equation 22-1

$$x = \frac{x'}{\sqrt{1 - \frac{v^2}{c^2}}}$$

This is often interpreted as a real contraction of space itself. Once again there are an infinite number of possible relative velocities, and there are an effectively infinite number of stars moving relative to space at different velocities. How could space itself be contracted over an infinite range of velocities in every direction simultaneously? Of course, it cannot. Such an idea is ridiculous. The underlying geometry of space is flat without contractions or curves.

Next we have to ask ourselves, does space know how to tell time? If we consider space without even the vacuum fluctuations, there is no timekeeping mechanism. Clocks are made of matter, and while they may have varying clock rates, there is no connection with the underlying space, if we take away the zeptons from consideration. As far as space is concerned, the only time is universal time. Time is not a relation to space; time is a relation to energy. Time dilation only occurs with respect photons or matter. More properly, time dilation needs to be grouped with energy transformations, not space dimensions.

Frame Transformations of Space
When we take away the ideas that space is contraction or that space knows how to tell time, we are left with frame transformations of space that are strictly Galilean. If there is a frame moving at a constant velocity **v** along

the x axis with position x' in the second frame of reference initially at the same position x in the first rest frame, the position at different times t_0 can be determined by Equation 22-2, where t_0 is universal time. On the remaining axes, $y = y'$ and $z = z'$, and as far as time goes, $t = t' = t_0$.

Equation 22-2

$$x = x' - vt_0$$

Equation 22-3

$$y = y'$$

Equation 22-4

$$z = z'$$

Equation 22-5

$$t = t' = t_0$$

Note that the frame transformations of space are not of any great utility; since they are not relative to anything we want to measure, such as matter, light or clocks. It is important, however, to set the framework and clear up misperceptions of contraction and time dilation of space.

Frame Transformations of Light
It would be nice if all frame transformations were truly as simple as the Galilean transformations of space; however, when it comes to transforming light and physical objects, we do see contraction of wavelengths and changes in clock rates. It is important to remember that even if these transformations are written in dimensions x and time t, they are not transformations of space.

We will find that it was the constancy of the speed of light assumption that led us down the wrong path all along. A constant speed of light assumption is only correct when there is no carrier for the photon, no æther. In the absence of an æther to carry light, the speed of light is presumed to be a fundamental property of light and not of æther. As soon as we recognize the existence of the zero-point field, we can no longer make the assumption that the speed of light is constant anywhere other than the zero-point field's rest frame. The elimination of æther from the theory of light caused us to go down an entirely incorrect path regarding frame transformations. It is not the case that the speed of light is constant and space contracted, but rather the speed of light changes and space is flat.

The speed of light is transmitted in the zero-point field rest frame, regardless of what frame it is emitted from. We can call the speed of light in the universal rest frame c_0. In every other reference frame moving at velocity V relative to the zero-point field rest frame, the speed of light changes in accordance with Equation 22-6. I will adopt here the convention of Larmor of using a capital V to designate the velocity relative to the universal rest frame. In 1900, he published an extensive development of frame transformations relative to a universal rest frame including applications to electromagnetic theory.[116] It is important to note that the velocity changes can alternatively be thought of being due to an increase in the London-van der Waals torque in the moving reference frame.

Equation 22-6

$$c' = c_o - V$$

This is simply illustrated in Figure 22-1, where there is a moving reference frame A moving at velocity V relative to the zero-point field rest frame. A photon emitted from rest frame A moves at a slower speed than the speed of light in the universal rest frame.

We can imagine how the velocity difference occurs in a simple way. If the zeptons have to rotate to reflect a frame moving at velocity V, and given a maximum degree of rotation of **180 degrees**, they will have rotated by an average amount of **180 x V/c_0 degrees**. The remaining amount of rotation they have left is **180 x (c_0-V)/c_0 degrees**. This gives an appearance of an increase in van der Waals torque from the perspective of the moving reference frame.

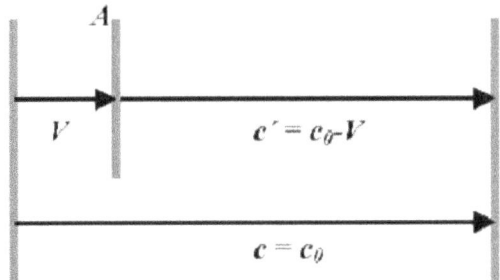

Figure 22-1 The lower arrow is a photon moving relative to the zero-point field rest frame at the speed of light **c_0** and the upper arrow is a photon emitted from rest frame **A** moving at velocity **V** with respect to the zero-point field rest frame.

Then if the beam of light is reflected and goes back the other way, the light still travels the speed of light c_0 relative to the universal rest frame, but now has the extra velocity of the rest frame as shown in Equation 22-7.

Equation 22-7

$$c' = c_0 + V$$

This leads to an important principle of relativity with respect to a universal frame of reference. The average two-way velocity of light in a roundtrip is c_0. This is fundamentally different than saying that the speed of light is constant between every observer. At the same time, it matches our observations since we only have one platform, Earth, where we can both emit and detect light; most of our experiments require a round trip.

Recall next the basic relationship between the speed of light c, the wavelength λ and frequency f as shown Equation 22-8. Note that f is used for frequency in this chapter to avoid confusion with velocity. From this relationship it is clear that within the scope of existing relativity theory, both the wavelength and frequency of light must be different in a reference frame other than the universal rest frame in order for the speed of light to be constant. The wavelengths would be shorter and the frequency lower.

Equation 22-8

$$c = f\lambda$$

On the other hand, since the one-way speed of light is lower for light emitted from a rest frame moving relative to the universal rest frame, it is not the case that both the frequency and wavelength must change. We can look at a similar image to what we used before, but this time the outer walls will equal one wavelength in the zero-point field rest frame. Rest frame A is moving at the same velocity V so it moves a distance that equates to the wavelength times the ratio of V/c_0. The wavelength of light emitted from a source on the moving rest frame is then shortened by that amount.

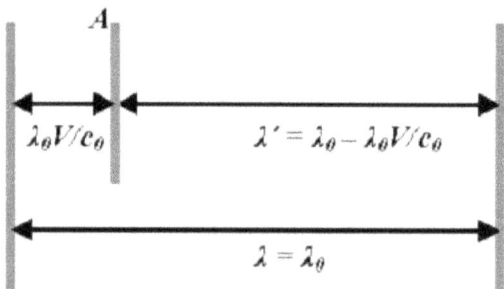

Figure 22-2 The lower double arrow is a wavelength in the zero-point field rest frame at c_0 and the upper arrow is a photon emitted from rest frame A moving at velocity V with respect to the zero-point field rest frame.

The frequency of the light as seen from an observer in the zero-point field rest frame, using a clock in the zero-point field rest frame, is the same for the photon emitted in his frame of reference and an identical photon emitted from rest frame **A**. If you look at Figure 22-1 again, it is simple to see that there are an identical number of wavelengths in the two photons. It is just that the wavelengths are shorter as viewed form the more compact appearing space of the moving reference frame.

The observer on reference frame **A** sees things differently. He sees a photon emitted from his frame of reference moving at the speed of light c_0. How does that happen? It happens because the clocks in his frame of reference run slow, making the frequency appear higher by the ratio c_0/V. This is opposite from the ratio applied to the wavelength, so when the frequency and wavelength are multiplied together as in Equation 22-8, it leaves c_0. This is in no way saying that the time of space is running slow or dilated, or even that such a thing could ever be logically be conceived to be a possibility; it is purely clock slowing. The clocks run slow because they loose energy to the zero-point field because they are moving relative to it.

These are how transformations have to been done for one-way transformations of light at low velocities **V**. Here we need to note that these one-way transformations can more generally be thought of as transformations of energy, where the kinetic and inertial energy $E=mv^2$ plus the photon energy as observed from the moving reference frame is equal to the photon energy in the universal rest frame. At higher velocities, where **V** is closer to **c,** more and more energy is required to get closer and closer to the speed of light. At such velocities, the relativistic **γ** term must be incorporated into the equations.

The Doppler Effect

The Doppler Effect on light is perhaps the single greatest piece of evidence we have that light undergoes a change in energy in accordance with frame transformation theory. It is also a one-way transformation like above. If a galaxy, for example, is moving away from us, the light we detect is said to be red-shifted. The wavelength becomes longer, and red is on the long wavelength side of the visible light spectrum. While the wavelength lengthens, the frequency is reduced, while at the same time the measured energy of the photons is lower. If a galaxy is moving toward us, the light is blue-shifted. Blue is on the short wavelength end of the visible light spectrum. Blue-shifted light has higher frequencies and higher energy than expected based on the emission source. This effect is called a Doppler Shift, because it is reminiscent of the Doppler Effect for sound.

The classical Doppler Shift in terms of wavelength λ is quite simple as shown in Equation 22-9. The e subscript designates the emitted wavelength, and the d subscript indicates the detected wavelength. The ratio between the differences in wavelengths divided by the emitted wavelength equals the velocity v between the emitter and detector divided by the speed of light c.

Equation 22-9

$$\frac{\Delta\lambda}{\lambda_e} = \frac{\lambda_d - \lambda_e}{\lambda_e} = \frac{v}{c}$$

This mathematical expression of the Doppler Effect of light matches all observations. All measured Doppler Shifts due to relative velocities are linearly proportional to that velocity. The shift in wavelength occurs due to the change in velocity relative to the universal rest frame. This shift can also be considered as a change in energy, which is required so that energy is conserved. The Doppler Effect is in complete accordance with light transmission relative to a zero-point field rest frame.

Note that there is a relativistic Doppler equation that can be used when a distant object is moving at or near the speed of light. This is necessary in order to correct for the additional energy needed as the speed of light is approached.

The Sagnac Effect

Another important piece of evidence we have about the speed of light with respect to reference frames is the Sagnac Effect. The Sagnac Effect occurs when the reference frame is rotating. It is the Sagnac Effect that is partially responsible for clocks in orbit running at different rates than they do on the Earth's surface. It is the Sagnac Effect that is likewise responsible for clocks carried on airplanes, or generally being transported around the globe, losing or gaining time. It is important to note that you will frequently find it written that the Sagnac Effect is due to Special Relativity. It is not. The Sagnac Effect requires a standard reference frame, in our case Earth, which Special Relativity does not allow for. The apparent velocity of light relative to a rotating emitter/detector is also not **c**. Note that the time rate change can be calculated relative to the Earth's non-rotating rest frame, without referring directly to the zero-point field rest frame.

The Sagnac Effect occurs because in a rotating frame of reference the effective speed of light includes the velocity due to rotation as shown in Figure 22-3. An emitter at the surface of a rotating sphere emits one photon in its direction of rotation and another in the opposite direction. We assume that light is somehow reflected so that it stays in orbit. The velocity of both photons is **c** with respect to the Earth non-rotating rest frame, but not with respect to the emitter/detector. As the sphere continues to rotate, the photon sent in the direction opposite from the rotation is detected first. From the perspective of the observer on the surface, it appears that this photon traveled at **c + v** where **v** is the velocity of rotation at the surface of the sphere. The other photon, which arrives later, appears to travel at the velocity $c - u$ from our observer's perspective, and so it arrives noticeably later.

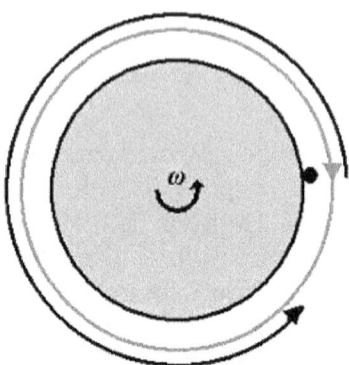

Figure 22-3 The Sagnac Effect shown with a gray rotating sphere in the center with a light emitter at the surface represented by the black dot. It simultaneously emits two photons, one in the direction of rotation and one in the opposite direction. The photon sent in the opposite direction is detected back at the emitter first.

As you can see from the $c \pm u$ coefficients, the Sagnac Effect behaves in a manner consistent with the Galilean frame transformations at low velocities. The speed of the photons does not change and even the Doppler Effect does not appear, as the one photon really does travel a shorter distance than the other. Why then is this effect hailed as a triumph for Special Relativity Theory? That goes back to Michelson, who thought this could be a test of whether or not there was an æther drag effect caused by the rotation of the Earth.[117] It was never intended as a test of Special Relativity, but of æther theory. Of course we know that there is no æther drag, so the Sagnac Effect proves nothing with respect to relativity theory or the existence of æther.

If instead of photons we send planes or satellites around the Earth, we can see the Sagnac Effect. The change in reference frame leads to a real change in total energy, which causes clocks on board to run slower or faster than they would at a fixed location on the Earth's surface. The time correction is made with Equation 22-10.

Equation 22-10

$$t' = t + x' \frac{v}{c^2}$$

In the above equation, x' is determined from Equation 22-2. This equation is identical to the one derived by Larmor in 1897.[118] With any clock in flight or orbit there is also a second time correction, one due to Newtonian gravity, which is discussed in the next chapter.

Transformations of Energy, Mass, and Time

There are three groups that need to be considered when analyzing frame transformations, and each must be addressed separately. The first is space, the second is light, and the third is energy. It is finally with respect to energy that we find out what really happens with physical rods and clocks. The energy of any matter in a rest frame that is moving with respect to the universal rest frame includes the energy of the matter in the universal rest frame, plus with the total kinetic and inertial energy it gains due to the velocity of the reference frame. That means we have the rest energy E_0 as expressed in Equation 22-11, where m_0 is the rest mass.

Equation 22-11

$$E_0 = m_0 c_0^2$$

Then we have the kinetic plus inertial energy E_{ki} in Equation 22-12 where each, kinetic and inertial, is responsible for half the energy due to the velocity of the reference frame.

Equation 22-12

$$E_{ki} = m_0 v^2$$

What we find when computing anything related to frame transformations is that it all comes back to energy conservation. If the wavelength of light changes, such as with the Doppler Effect, it can be computed using energy conservation principles. If the time rate of a physical clock changes, it can be computed using conservation of energy principles. The relativistic transformation for energy is shown in Equation 22-13.

Equation 22-13

$$E' = \frac{E_0}{\sqrt{1 - \frac{v^2}{c_0^2}}}$$

A related term is something called relativistic mass. It is sometimes useful to treat mass as being proportional to energy, so one can come up with a relativistic mass term as shown in Equation 22-14.

Equation 22-14

$$m' = \frac{m_0}{\sqrt{1 - \frac{v^2}{c_0^2}}}$$

In the model for mass presented in Chapter 16, mass is equivalent to the vacuum fluctuation energy excluded by a spherical shell the size of the charge radius of the proton and electron. When a particle, or specifically a spherical shell, is moving with respect to the zero-point field rest frame, the excluded wavelengths will be more energetic due to the additional kinetic/inertial energy. The particle does not have to physically contract in order to gain energy or effective mass-energy.

Atomic orbits and molecular spacing are affected by the change in total energy of the particles, but not with respect to a single dimension. Any contraction at the atomic level due to energy occurs in three physical dimensions, so that any contraction will be much smaller than would be expected due to Lorentzian length contraction in one dimension. This is why physical rod contraction has never been seen experimentally. We also must note that the true rest mass of a particle or any physical object can only be measured with respect to the universal rest frame. Physicists really do need to eliminate the term "rod contraction" from their vernacular, at least with respect to being equivalent to photon wavelength contraction.

The final piece of the frame transformation puzzle is the time rate change of clocks. As stated before, this phenomenon is not a time dilation of space, but simply a change in clock rate due to the kinetic energy due to the velocity of a physical clock relative to the universal rest frame. The equation for the time rate change is as shown in Equation 22-15.

Equation 22-15

$$\Delta t_0 = \frac{\Delta t'}{\sqrt{1 - \frac{v^2}{c_0^2}}}$$

A first order approximation of the above equation is what we more commonly see in computations we make on Earth, such as for GPS clocks. We have found that first order approximations for virtually all relativistic equations are close enough for what we need to compute. The basic time rate change equation was shown in Equation 22-10.

If we consider atomic clocks based on the vibration of atoms, it is known that the vibration frequency of atoms change with respect to the velocity relative to the zero-point field rest frame. This leads to a real change in the rate of time as measured by atomic clocks. It is only when we study physical clocks that we see a time rate change as it is not a property of space. Time rate change is ultimately due to the change in energy of the matter that the clock is made from.

While we will not go into any detail on how the transformations of energy affect the Electro-Matter Force, it is important to note that like light, the electro-matter forces propagate through the universal rest frame. Larmor began the work of developing a theory of electricity and magnetism including frame transformations with respect to a stationary æther.[116, 118] A complete Electro-Matter Force Theory including transformations with respect to the zero-point field rest frame is something that will require additional development.

Selleri Transformations

As it turns out, Franco Selleri has already derived the proper frame transformations relative to a standard æther rest frame.[119] He does retain the convention of performing the transformations of photon wavelengths and frequencies as transformations of space and time, respectively. He refers to "rod length" contraction rather than sticking with the wavelength contraction concept, but otherwise his work appears to be complete.

Equations 22-16, 22-17, 22-18 & 22-19

$$x = \frac{x_0 - \beta c t_0}{\sqrt{1 - \beta^2}}$$

$$y = y_0$$

$$z = z_0$$

$$t = \sqrt{1 - \beta^2}\, t_0 + e_1(x_0 - \beta c t_0)$$

His equations as presented in his referenced paper as Equation 29 are shown below as Equations 22-16, 22-17, 22-18, and 22-19. Note that the term beta $\beta = v^2/c^2$ is per the traditional shorthand notation. In his equations, the universal rest frame is noted by the $\mathbf{0}$ subscript. The remaining coefficient $\mathbf{e_1}$ is shown in Equation 22-20.

Equation 22-20

$$e_1 = -\frac{1}{c}\frac{\beta}{\sqrt{1-\beta^2}}$$

Conclusions

Zero-point field theory is an æther theory, and since light travels through the æther, the speed of light is relative to the æther rest frame. All matter travels through the æther as well, and its properties of energy, mass, and time rate must be adjusted accordingly. Photons only travel the speed of light in the zero-point field rest frame and $\mathbf{c \pm u}$ in any other frame moving at velocities $\mathbf{v}$, with the two-way round trip of a light signal averaging out to $\mathbf{c}$. Einstein's contributions to the postulates of relativity were entirely unnecessary, and entirely wrong, as was his idea that the æther was unnecessary.

To summarize in simple list form, we have the following:

54) Light is transmitted in the zero-point field rest frame
55) The second postulate of Special Relativity that light travels at velocity $\mathbf{c}$ with respect to any observer is incorrect
56) In a rest frame moving with respect to the zero-point field rest frame the speed of light is slower for light emitted in the direction of the frame velocity and faster in the direction opposite the frame velocity by $\mathbf{c \pm u}$

57) The average two-way roundtrip speed of light is **c** when measured in a given reference frame
58) Space does not undergo length contraction
59) Space does not undergo time dilation
60) Space by itself can be transformed in a strictly Galilean manner
61) Photon wavelengths are contracted
62) Length contraction should be more properly expressed as photon wavelength contraction
63) Photon frequencies are lengthened
64) With respect to light, time dilation should more properly be expressed as photon frequency lengthening
65) The Doppler effect is required due to energy conservation
66) Measured time rate changes due to the Sagnac Effect are due to energy conservation with respect to physical clock mechanisms
67) The true rest mass of a body must be measured in the zero-point field rest frame
68) Rod contraction is due to a change in energy and is in respect to the entire volume, not a single dimension
69) Electro-Matter Force translations must be made with respect to the zero-point field rest frame

[114] H. A. Lorentz, The Theory of Electrons, 2nd Edition , B. G. Teubner, Leipzig; G. E. Stechert & Co., New York, 1916. (based on lectures given in 1906).

[115] J. C Maxwell, 1865, A Dynamical Theory of the Electromagnetic Field, Vol. 1, pp. 579 – 580, Dover Publications, New York, 1890.p579-580

[116] J. Larmor, Aether and Matter, Cambridge University Press, 1900.

[117] A.A. Michelson, "Relative Motion of Earth and Aether". Philosophical Magazine 8 (48): 716–719, 1904.

[118] J. Larmor, "On a Dynamical Theory of the Electric and Luminiferous Medium, Part 3, Relations with material media", Phil. Trans. Roy. Soc. 190: 205–300, 1897. doi:10.1098/rsta.1897.0020.

[119] F. Selleri, Remarks on the transformations of space and time, APEIRON, Vol. 4 No. 4, Oct 1997.

Chapter 23: Gravity, Energy, and Light

Recapitulating, we may say that according to the General Theory of Relativity space is endowed with physical qualities; in this sense, therefore, there exists an Aether. According to the General Theory of Relativity space without Aether is unthinkable; for in such space there not only would be no propagation of light, but also no possibility of existence for standards of space and time (measuring-rods and clocks), nor therefore any space-time intervals in the physical sense.[120]

Albert Einstein, 1920

Einstein, Confusing Physicists for Over a Century

There are four remaining types of observations, which are considered to be proofs of General Relativity that involve interactions between gravity and energy and/or light. The first is the change in the time rate of clocks in a gravitational field, the second is gravitational redshift, the third is the Shapiro Delay, and the fourth is the bending of light. The first two deal primarily with differences in energy due to tangential or radial motion relative to the Sun or Earth. The second two involve light moving tangentially to the Sun, which includes both the radial and tangential components, thus multiplying the gravitational effects.

Of the above phenomena, the one on the most solid footing experimentally is the time rate change of clocks, as a "gravitational" correction is required for clocks used in GPS satellites. Parentheses were used above because as with many things relativistic, the transformation is required due to a change in the state of energy due to the Matter Force and long-range van der Waals gravity, rather than Newtonian gravity.

In the traditional theory, clocks are said to run slower when they are in a region of space with greater acceleration due to Newtonian gravity. This is said to come about because of the Equivalence Principle, which roughly states that light interacts with gravity in the same way that mass does. This clock rate speeding up as the distance from Earth increases requires that the clocks on GPS satellites must be set to run slower than the equivalent clock on Earth. At the same time they are adjusted to run faster due to the velocity with respect to Earth due to the Sagnac Effect. The time rate change due to "gravity" is, however, much greater.

The basic equation used for this gravitational time correction is Equation 23-1, where G is Newton's gravitational constant, M is the mass of the Earth, r is the radial distance between the clock and observer, and c is as the speed of light. Note that this equation does not adhere to Special Relativity because it is necessary to introduce the concept of proper time (t_0), the time in a standard frame of reference. Special Relativity does not allow for a standard reference frame, a point generally ignored by proponents of Special Relativity. Without a standard reference frame it would not be possible to compute a time rate change.

Equation 23-1

$$t_0 = t'\sqrt{1 - \frac{2GM}{rc^2}}$$

We can readily recognize within the scope of zero-point field theory what is really going on. The clock rate change is not due to a time dilation effect or acceleration, but rather the London-van der Waals torque of the vacuum is increased in the presence of nearby large bodies of matter. The speed of rotation of the zeptons in space is slowed due to the polar Matter Force. This additional torque causes photon wavelengths to be longer and the frequencies lower. In turn it causes physical clocks to run slower than they do farther from Earth. It is important to note that while

236

Einstein derived this idea from the concept of acceleration and the Equivalence Principle, it is more properly treated as a Matter Force and energy conservation problem. It is also not necessary to include the square root in the equation, as a simplified first order approximation is usually sufficient.

To derive the formula from an energy perspective, we can start with the Newtonian gravitational potential energy shown in Equation 23-2. Note that for simplicity of the discussion, this will not be broken into its two principle components, the Matter Force and the long-range van der Waals Force. The potential term includes **GM** and the distance **r** between the point where the potential is being measured and the center of mass. The negative sign means it is attractive.

Equation 23-2

$$E_p = -\frac{GM}{r}$$

Alternatively, we can look at the potential energy per rest mass energy. Recall that the energy **E = mc²**, or alternatively **m = E/c²** where **m** is what needs to be substituted into the above equation. By dividing the above equation by **c²** we can then calculate the force on the photon by multiplying by the photon's energy. This allows us to use the Equivalence Principle to say that photon energy is equivalent to mass-energy, and computations can be made accordingly. Then we get Equation 23-3.

Equation 23-3

$$E = -\frac{GM}{rc^2}$$

Plugging that in, we get a time correction factor as shown in Equation 23-4. A square root is added around the subtracted terms to get the relativistic form, but can be left off when the first approximation is sufficient, which it usually is.

Equation 23-4

$$t_o = t'\sqrt{\left(1 - \frac{GM}{rc^2}\right)}$$

It is important to note that the Equivalence Principle often appears in literature as being part of General Relativity theory, but is more correctly viewed as an entirely separate concept, as it does not overlap with other postulates of General Relativity. Evidence for the Equivalence Principle in no way supports General Relativity, just the Equivalence Principle. Beyond that, there is contention about whether or not light truly is affected by gravity, whether the Equivalence Principle is real. In the zero-point field model, the slowing of clocks in a gravitational field is due to the slowing of rotation of the zeptons due to the presence of mass. The Equivalence Principle in its traditional sense is unnecessary to account for clock slowing, as it should be replaced by more precisely correct ways of describing the force interaction.

In summary, what we have found is that the time rate corrections for clocks can be made strictly on the basis of the principles of the Matter Force and energy conservation, and in particular the slowing of the rotation of zeptons in the zero-point field. This correction is required when a body is moving tangentially to the Earth's surface, or any other body of matter. This slowing leads to an increase in the London-van der Waals torque, which leads to longer photon wavelengths and slower clocks.

Gravitational Redshift

While gravitational redshift is usually credited as being due to General Relativity, it can be simply considered to be a natural consequence of two more basic concepts: conservation of energy and the Equivalence Principle. Any theory that combines those two concepts will automatically have gravitational redshift. General Relativity is not required for there to be gravitational redshift, so conversely gravitational redshift is not a proof of General Relativity theory. As before with clocks, this has nothing to do with relativity, but has falsely been trumpeted as a triumph of relativistic theory.

From the energy perspective, gravitational redshift comes about because it takes energy for a photon to move in opposition to the gravitational potential field in the radial direction. The photon becomes less energetic as it moves away from the star that emitted it. In the process of becoming less energetic, the wavelength becomes longer. For visible light, it shifts toward the red end of the visible light spectrum, giving us the name redshift.

Equation 23-5 shows the equation for the shift in wavelength when the position of the emitter e is at a point r from the center of mass and the detector d is effectively at infinity. G is the gravitational constant, M is the mass of the star and c is the speed of light.

Equation 23-5

$$\frac{\lambda_d - \lambda_e}{\lambda_e} = \frac{GM}{c^2 r}$$

To understand this interaction in terms of the zero-point field, we can compare it to changes in the velocity of light relative to a moving rest frame with varying velocity v. In a moving rest frame, if light is emitted in the direction of the rest frame motion, it goes slower than the speed of light by c - v. The wavelength is contracted relative to the normal wavelength at c. The effective torque of the vacuum with respect to the moving frame is higher than in the universal rest frame. Then if we imagine that the relative frame velocity becomes slower $v_1 < v$, the light being emitted now travels $c - v_1$, which is faster than $c - v$. Since the speed of light is faster, the wavelength is longer, while at the same time the effective torque of the vacuum is reduced. As the moving rest frame's velocity continues to decrease in this manner, the light becomes increasingly redshifted as the speed of light increases. At the same time, the torque of the vacuum as seen in the light's frame of reference is decreasing. This is analogous to what is happening in the gravitational field. To put it more plainly, photons accelerate as they leave a gravitational field in the radial direction, and as they speed up their wavelengths become longer.

While gravitational redshift is frequently and erroneously said to be a consequence of General Relativity, it is actually just a consequence of the conservation of energy and the slowing of the speed of light in the vicinity of matter.

Shapiro Delay

In 1964 Irwin Shapiro proposed a forth test of gravity to add to Einstein's first three, which came to be known as the Shapiro Delay.[121] He proposed that radar signals could be bounced off a planet or satellite on the opposite side of the sun, and that there would be a delay of ~**200 microseconds** due to the effects of gravity when the radar signal passed near the Sun, versus when the Sun was farther from the signal path. He went on to demonstrate the effect a few years later and measured a delay of ~ **160 microseconds** when bouncing signals off the planet Mercury.[122] An Illustration of the Shapiro Delay is shown in Figure 23-1.

Shapiro repeated the experiment in 1977 using radio signals from the Viking Landers on Mars.[123,124] He obtained much more precise results than reported in the previous experiment after eliminating 10% of the data, which varied from the expected result by as much as

10%, or **~20 microseconds**.[124] There was also the difficulty that the solar corona was responsible for up to **100 microseconds** of delay, which had to be corrected for.[123]

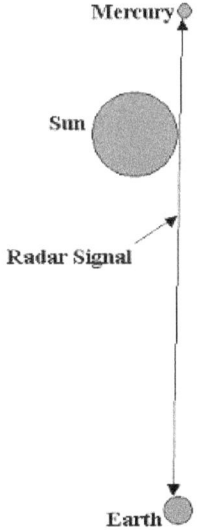

Figure 23-1 The Shapiro Delay experiment. A radar signal from Earth is bounced off Mercury and returns to Earth, grazing the Sun each way. The bending of the radar signal is not shown.

The measurement was quite a challenge, and perhaps the precision of the measurement is overstated, but the Viking measurements are considered to be conclusive proof of the delay in agreement with calculations due to General Relativity.

One might think that the delay is due to the bending of light, but the **1.75 degree** bend, which we will discuss next, can only be responsible for ~ **10 microseconds** of the delay.[125] The remaining **~240 microsecond** delay to and from Mars can only be accounted for within the scope of General Relativity theory as space contraction and space-time dilation to the tune of **~69 kilometers**. Of course it has never been an issue of space contraction, but rather the speed of light being reduced near bodies of matter. This is the same conclusion, the speed of light changing, reached by Paul Marmet who did an excellent job cataloging the difficulties involved with measuring the Shapiro Delay and the bending of light.[125] A word of caution though, as he was perhaps a little overzealous in his attempts to discredit all the relevant theories.

While there certainly should be doubt cast on some of the experiments purported to support the magnitude of the time delay and bending of light predicted under General Relativity theory, the position some scientists take that there is no delay or bending effect is an impossible one, due to the known time rate change of clocks. The time rate change of clocks in a gravitational field demands that there is at least some delay, some bending of light. The only question that might be open to debate is Einstein's factor of two. Even so, the data strongly suggests that the magnitude of the delay and bending he predicted are correct, which leads us with only one course of action: to understand the mechanism without resorting to space contraction and time dilation.

To begin that analysis we can examine the change in the speed of light, which can be calculated using Equation 23-6, note the factor of **2** relative to the correction terms in Equation 23-4 and 23-5 above.

Equation 23-6

$$c' = c\left(1 - \frac{2GM}{rc^2}\right)$$

Within the scope of General Relativity, we are told that half of the delay is due to space contraction and the other half is due to space time dilation, and that Einstein's initial error occurred because he initially only took time dilation into account. From the previous sections we can see that gravitational redshift gives us a wavelength contraction effect, while the torque that gives us the time rate change of clocks due to gravity also gives us a time delay. Slowing with respect to gravitational redshift is the radial motion component, while slowing due to London-van der Waals torque near matter is the tangential component. Combining the two gives us our factor of two, so that the speed of light changes in accordance with Equation 23-6.

Half of the speed of light change is due to the identical mechanism used to explain the gravitational redshift, the deceleration and acceleration of light as it approaches and leaves the vicinity of the Sun. The other half is due to the increase in the torque of the vacuum in the vicinity of nearby matter due to polar matter forces. As a photon travels tangentially to a large body of matter, the photon wavefront surrounding the central zepton extends indefinitely out into space through a non-uniform matter field - or more to the point, a zero-point field where the London-van der Waals torque of the vacuum is not uniform. This causes light to slow in a manner consistent with the time rate change of clocks. Note in particular that since the photons involved with gravitational redshift are not traveling tangentially but in the radial direction, those photons are not subject to this component of the Matter Force. The net effect is that both deceleration-acceleration of light and van der Waals torque of the vacuum affect photons when they arc in thc vicinity of matter. They combine to give us the same factor of two originally predicted by Einstein. The theory of General Relativity is unnecessary.

Stepping back a bit in time, if we imagine that we know nothing of General Relativity theory and we found experimental evidence that light is delayed as a light signal passes the sun, our first inclination would be to say, oh that is interesting, the Sun causes light to slow down. Space can still be presumed to be flat, while light slows down in the presence of acceleration due to "gravity." Einstein on the other hand in his theory of General Relativity says that no, the speed of light does not slow down, but instead space-time is contracted, time dilated and curved. So we have a choice between a nice geometrically well-behaved flat space, where light slows down near a large body of matter, or a constant speed of light with contracted, time dilated and curved space-time. Einstein chose poorly.

The Bending of Light Around the Sun

As previously mentioned, the next traditional test is the bending of light by the Sun. In this case, there is expected to be light bending due to Newtonian gravity equivalent to 0.875 arc seconds based on the Equivalence Principle, but Einstein changed his mind and added an additional factor due to length contraction, predicting an angle of 1.75 arc seconds. The bending of light is illustrated in Figure 23-2 where a beam of light from a distant star is bent as it passes nearby the Sun and then is detected at Earth. This, interestingly, makes it appear to the casual observer that the distant star is located somewhere, other than its real location.

Figure 23-2 A ray of light from a distant star bends as it passes near the Sun.

Equation 23-7 shows the basic formula for computing the angle based on the Newtonian gravitational constant.

244

Equation 23-7

$$\theta = \frac{4GM}{rc^2}$$

Within the scope of the Matter Force Theory this can also be described as a two-part process that is identical to Shapiro Delay, except in this case the slowing of light near matter additionally causes the light to bend. The clock slowing or frequency change due to the tangential motion is responsible for half the bending, while slowing due to the radial component is responsible for the other half.

Conclusion

A lot can be written about these four tests of General Relativity, but those are the basics. Clocks in a gravitational well run slower because not only the frequency of light, but the frequency of all matter in the gravitational well becomes longer. It is more appropriate to think of it as the clock running slower in the gravitational well due to the local increase in the London-van der Waals torque due to the local presence of matter. This phenomenon is not a physical time dilation of space. Gravitational redshift on the other hand is simply a case of energy being expended to escape the gravitational potential well. It is merely a consequence of energy conservation, the loss of photon energy as it accelerates out of a gravitational potential well.

The two tests where light passes near the Sun, the Shapiro Delay and bending of light, are likewise simply explained as being half due to deceleration-acceleration in the radial direction, which in turn can be thought of as being due to energy conservation. The other half is due to the non-uniform matter field of the Sun introducing an additional torque causing a time delay as the photon moves tangentially to the Sun. As with General Relativity, two somewhat different effects are responsible for the total amount of delay or bending, but they can be accounted for as a zero-point field interaction, which does not require that space-time be length contracted, time dilated, or curved. Einstein chose poorly. Space is flat, and the speed of light is reduced near matter.

That leaves us with four new key points:

70) The time rate change of clocks is due to a change in London-van der Waals torque with respect to the presence of local matter
71) Gravitational redshift is due to a decrease in energy as a photon accelerates away from a star
72) The Shapiro Delay is due to the reduction in light velocity due to photon deceleration/acceleration and increased London-van der Waals torque
73) The bending of light is due to the reduction in light velocity due to photon deceleration/acceleration and increased London-van der Waals torque

[120] A. Einstein, "Æther and the Theory of Relativity," Address delivered on May 5th, 1920, at the University of Leyden, Germany

[121] I. I. Shapiro, "Fourth Test of General Relativity". Physical Review Letters 13 (26): 789–791. Bibcode 1964PhRvL..13..789S. doi:10.1103/PhysRevLett.13.789, 1964.

[122] I. I. Shapiro, et. al. "Fourth Test of General Relativity: Preliminary Results". Physical Review Letters 20 (22): 1265–1269. Bibcode 1968 PhRvL..20.1265S. doi:10.1103/PhysRevLett.20.1265.

[123] I. I. Shapiro, et. al., The Viking Relativity Experiment, Journal of Geophysical Physics, 82, 28, p. 4329-4334, 1977.

[124] R.D. Reasonberg, I.L. Shapiro, et. al., Viking relativity experiment: verification of signal retardation due to solar gravity, The Astrophysical Journal, 234:L219-221, Dec 15, 1979.

[125] P. Marmet, and C. Couture, Relativistic Deflection of Light Near the Sun Using Radio Signals and Visible Light, presented at the University of Storrs, Connecticut, at the meeting: "The New Natural Philosophy: An Introduction to the 21st Century Physics and Cosmology" Storrs, 4-8 June 2000. http://www.newtonphysics.on.ca/eclipse/

Chapter 24: Standard Tests of Gravitational Theory

> *For large densities of field and of matter, the field equations and even the field variables which enter into them will have no real significance. One may not therefore assume the validity of the equations for very high density of field and of matter, and one may not conclude that the 'beginning of the expansion' must mean a singularity in the mathematical sense. All we have to realize is that the equations may not be continued over such regions.*[126]
>
> Albert Einstein, 1956

First, The Zero-Point Universe Tests

When Einstein developed General Relativity, he proposed three tests that any competing theory would have to match in order to be seriously considered. Those tests were the precession of the perihelion of Mercury, gravitational redshift and the bending of light by the Sun. Since then those tests have become standard for any new theory of gravity, and other tests have been added along the way. To follow tradition, we shall cover them here. But first I will begin with a few tests of my own. Fair is fair.

1. Does the Theory Factor in Zero-Point Vacuum Fluctuations?

General Relativity fails this test, since vacuum fluctuations are not a consideration in any of its formulas or physical explanations. There is no part of the theory that explains how zeptons participate in the contracting and bending of space. If the energy of the vacuum is recognized as part of the total energy of space according to General Relativity, then the universe would be compressed to a point, so the energy of the vacuum is simply ignored. To make it worse, physicists are two-faced about it, accepting vacuum fluctuations in such things as quantum field theory, the Casimir Effect, and

Hawking Radiation, while excluding zero-point energy from General Relativity theory.

The long-range van der Waals and Matter Force theories on the other hand incorporate zero-point vacuum fluctuations at their very foundation.

General Relativity: **Failed**
Long-range van der Waals and Matter Forces: **Passed**

2. Does the Theory Require a Prior Structure?

A force law must not impose any prior structure on space or space-time. It must fundamentally be uniformly isotropic and geometrically flat. In other words, it must be the same everywhere and adhere to basic Euclidean geometry. General Relativity fails this test, as it requires that space-time be length contracted, time dilated and curved. Because of the prohibition within the theory of faster than light communication, space must know where the mass is in advance without any messages from matter throughout the universe being transmitted to each point in space. Even if you do allow for faster than light communication, you still have the problem of how could every bit of matter in the universe communicate with every region of space in the universe nearly instantaneously so that space somehow knows how much curvature it should have. General Relativity cannot escape the prior structure problem.

The long-range van der Waals Force and Matter Force theories pass this test as they fundamentally start with a uniformly isotropic zero-point field throughout a geometrically flat space.

General Relativity: **Failed**
Long-range van der Waals and Matter Forces: **Passed**

3. Is the Force the Correct Magnitude?

This can also be referred to as the missing mass problem. In General Relativity theory, it was known initially that there was approximately 10% of the mass required for certain astronomical bodies, such as larger galaxies, to have formed. Over time that number has increased to about 30% as various forms of the so-called "Dark Matter" have been identified. It should not matter whether you get an answer 10% correct or 30% correct, either must be deemed to be a failure.

The long-range van der Waals and Matter Force theories pass this test with assistance of Lorentz forces due to mattermagnetism. This additional inward force makes up for the missing mass.

General Relativity: **Failed**
Long-range van der Waals and Matter Forces: **Passed**

4. Does the Theory Introduce Space-Time Singularities?

Singularities can be fascinating when watching or reading science fiction, but they have no place in real science. Singularities are mathematical anomalies, which should not occur in an accurate model of the universe. The basic model of the universe must ideally start out and remain geometrically flat. General Relativity fails this test as it contains singularities. Einstein's answer in the quote at the beginning of the chapter is to simply ignore those parts of the equations. Better yet, we should ignore the theory and find another one, which does not introduce singularities in the first place.

The long-range van der Waals and Matter Force theories pass this test as they have uniformly and geometrically flat space-time without singularities, since there is no contraction or curvature of space-time issue.

General Relativity: **Failed**
Long-range van der Waals and Matter Forces: **Passed**

5. Does the Theory Agree with Conservation of Energy-Momentum?

Because General Relativity deals with gravity as curvature of space-time rather than as a force, there is not a direct way to compute force/energy/momentum for bodies moving along the geometric curvature of space. This leads to a problem where the energy and momentum in relation to gravity must be backed in due to Newtonian relations. In cases where the Newtonian and General Relativistic models do not agree, there is no way to tell if energy and momentum are conserved.

The Electro-Matter Force Theory on the other hand does agree with conservation of energy and momentum.

General Relativity: **Failed**
Long-range van der Waals and Matter Forces: **Passed**

6. Can the Theory Account for Superposition of an Attractive Force with a Repulsive Force?

This question comes about because the expansion of the universe is accelerating; therefore, there must be a force responsible for that acceleration. That means that we have two forces, one attractive and one repulsive. Classical strength gravity, either Newtonian or General Relativity, is a superposition of these two forces. How can a fundamental force, which is determined by curvature of space due to the presence of matter, be derived as the superposition of two forces? It cannot. Someone might make the argument that one force or the other is due to curvature of space, but not both simultaneously.

On the other hand, matter repulsion and the long-range range van der Waals Force both exist with no theoretical conflict.

General Relativity: **Failed**
Long-range van der Waals and Matter Forces: **Passed**

7. Does the Theory Account for Spiral Galaxy Formations?

General Relativity does not. It fails both in terms of not providing the force strength necessary to retain the most distant stars and not have a force mechanism that explains the spiral arm formation.

As discussed in a previous chapter, the long-range van der Waals Force and Matter Force theories pass this test. It has the additional force strength due to the mattermagnetic Lorentz forces, and the banding is caused because adjacent stars moving side-by-side are attracted due to mattermagnetic forces.

General Relativity: **Failed**
Long-range van der Waals and Matter Forces: **Passed**

8. Does the Theory Account for Tidal Forces?

General Relativity does not. It does not contain a long-range force that affects the rotation of distant bodies. There are the Lense-Thirring Effect and de Sitter Effect, which are related to rotation, but they have not been applied to long-range tidal forces. They will be discussed separately as one of the classical tests.

The mattermagnetic field produced by the rotation of one body affects the rotation of nearby bodies leading to the "tidal forces." If the two bodies are initially rotating in the same direction, matter forces causes the rotation of both bodies to slow.

General Relativity: **Failed**
Long-range van der Waals and Matter Forces: **Passed**

9. Does the Theory Account for the Anomalous Acceleration in the Solar System?

General Relativity does not contain a term that accounts for a slight increase in gravity as we get further away from the Sun while in our solar system.

Within the scope of the Electro-Matter Force Theory, the anomalous acceleration can easily be described a mattermagnetic Lorentz force due to the planets and asteroids in orbit around the sun.

General Relativity: **Failed**
Long-range van der Waals and Matter Forces: **Passed**

10. Can the Force be Unified with Electromagnetic theory into a Single Theory that Explains Both?

All efforts to unify General Relativity with electromagnetic theory have failed. Einstein spent much of his later life in the attempt, and the master manipulator was unable to make it work. Countless others have failed as well, and many physicists feel, correctly, that the two can never be joined.

The long-range van der Waals theory for gravity is on the other hand a simple extension of electro-matter theory. The Coulomb and London-van der Waals Forces are well established. The combination of electromagnetic and mattermagnetic forces into a single unified force is obviously simple as the two are charge based theories that are well characterized by classical physics.

General Relativity: **Failed**
Long-range van der Waals and Matter Forces: **Passed**

The Classic Tests
Do you think 10 was enough? Yes, 10 is probably enough to prove the point. Any one of those failures should be sufficient reason to dismiss General Relativity from serious consideration on its own. Nonetheless, we shall continue on the assigned task and compare the results of the classic tests of gravitational theories.

A. The Precession of the Perihelion of Mercury
This was already covered in a prior chapter. The General Relativistic solution requires four parts time dilation, minus two parts space contraction, plus one part relativistic mass increase. That smacks of a numerologically derived answer. Throw in the time dilation and length contraction, which we now know are fictions, and all we can conclude is that General Relativity fails to account for the additional 43 arcseconds per century in a theoretically sound manner. Nonetheless, we will give it credit as it is said to be a success within the scope of the theory.

The mattermagnetic force due to the rotation of the Sun on the other hand produces a previously unaccounted for classically acting force that is tangential to the direction of the velocity of Mercury and causes a small amount of additional precession. Even if we were to say that both theories solved the problem, if you had to chose between the two, a classically acting magnetic-like force is much preferable to one that requires a mixed bag of time dilation and length contraction.

General Relativity: **Passed**
Long-range van der Waals and Matter Forces: **Passed**

B. Gravitational Redshift

As discussed in the previous chapter, gravitational redshift can be considered a natural consequence of conservation of energy and mass-energy equivalence, the Equivalence Principle. Any theory that combines those two concepts will have gravitational redshift. It is not a test of General Relativity at all, as conservation of energy is expected to be true in any reasonable theory and the Equivalence Principle is acceptable to most physicists. In some places you will even see the Equivalence Principle by itself erroneously listed as a test of General Relativity.

What is different in the two theories of gravity is the physical interpretation. In the case of General Relativity, the interpretation is that in a gravitational well, length is contracted, such that the photon wavelength increases as the photon moves out of the well. This effectively increases the frequency and energy while reducing the wavelength of the photon, which is interpreted as length contraction. Within the scope of the theory, we can call this test passed, although the physical interpretation of time dilation and length contraction of space-time is fundamentally incorrect.

Within the scope of the Electro-Matter Force Theory, the cause of the shortening of the wavelength in a high gravity field is due to energy conservation and the lengthening of photon wavelengths as they accelerate out of a gravitational field. As the photon gets farther from the massive object it is redshifted as the speed of the light increases.

General Relativity: **Passed**
Long-range van der Waals and Matter Forces: **Passed**

C. The Bending of Light Around the Sun

In General Relativity theory, the bending of light is due in part to time dilation and the other part due to length contraction. The first half can also be considered the Newtonian part due to the acceleration due to gravity. In General Relativity, space-time bends in the vicinity of a body of matter. Light simply follows the space-time curvature. As before we can say that General Relativity passes the test within the scope of the theory while otherwise failing because curvature of space-time is nonsense.

In the long-range van der Waals and Matter Force theories, the bending of light is still a two-part process. The first part is bending due to the change in the velocity, due to the deceleration and acceleration of light due to energy conservation in the presence of matter. This is the radial component. The second part is due to the torque of the vacuum increasing with respect to a photon moving tangentially to a body of matter at the speed of light.

General Relativity: **Passed**
Long-range van der Waals and Matter Forces: **Passed**

D. The Shapiro Delay

Both arguments are essentially identical to the bending of light case.

General Relativity: **Passed**
Long-range van der Waals and Matter Forces: **Passed**

E. The Lense-Thirring Effect and de Sitter Effect

The last test we will consider, one that has been added to the list, is the Lense-Thirring Effect, which is often called frame-dragging. This has been characterized as a gravitomagnetic-like effect due to the rotation of a large body of mass. This is said to be due to General Relativity and leads to precession. It is sometimes discussed together with the de Sitter Effect or de Sitter Precession, commonly called the Geodetic Effect. The de Sitter Effect is an effect on bodies in orbit around a non-rotating body. It is also a gravitomagnetic-like force. They are frequently discussed together, because both forces are present at the same time, although the Lense-Thirring Effect is smaller in magnitude.

These are both of course due to the mattermagnetic force and have nothing to do with gravity, but we can be generous here and give General Relativity a pass.

General Relativity: **Passed**
Long-range van der Waals and Matter Forces: **Passed**

Conclusion
There are some other tests, but those five are representative and the most important. The previous 10, however, make the greatest distinction between the two theories and show that the long-range van der Waals and Matter Force theories of gravity are superior in every respect. In almost every case the claimed property of General Relativity is actually related to the previously unidentified or ignored Matter Force. One can only wonder how different physics would be if Maxwell had used his vortex theory to describe Newtonian mechanics back in 1870, allowing us to avoid all this General Relativity nonsense.

The key points of this chapter are that:

74) The theory of General Relativity fails numerous tests
75) The long-range van der Waals and Matter Force theories pass all the standard tests
76) Gravity as a long-range van der Waals Force is a part of the Electro-Matter force

[126] A. Einstein, The Meaning of Relativity, Princeton University Press p. 129, 1956.

Chapter 25: The Strong Nuclear Force

The transition of a heavy particle from neutron state to proton state is not always accompanied by the emission of light particles, i.e., a neutrino and an electron, but the energy liberated by the transition is taken up sometimes by another heavy particle, which in turn will be transformed from proton state into neutron state. If the probability of occurrence of the latter process is much larger than that of the former, the interaction between the neutron and the proton will be much larger than in the case of Fermi, whereas the probability of emission of light particles is not affected essentially.[127]

Hideki Yukawa, 1935

The History of the Strong Force

The force that binds nucleons, protons, and neutrons together in the nucleus of an atom has been a puzzle to physicists since the nucleus and proton were discovered. At distances greater than a few femtometers (fm = 10^{-15} m), two protons are repelled by the Coulomb charge, being that they are both positively charged. When they are somehow forced together such that they are within about **2 fm** of each other, they become attracted. This attractive force is known to be **~100** times stronger than the Coulomb repulsion, making it by far the strongest of the four fundamental forces in the Standard Model. The force then mysteriously becomes very weak again as the nucleons approach within about **0.7 fm**, with the net force even becoming repulsive.

To get an idea how close this is, the 2010 CODATA value for the charge radius of the proton is **0.8775(51) x 10^{-15} m**, so the range from **0.7** to **2.0 fm** is equivalent to **0.8** to **2.2** times the proton charge radius, or **0.4** to **1.1** times the proton diameter. We might immediately say that that sounds like ideal conditions for the van der

Waals Force; however, the van der Waals Force is normally so weak that that idea has historically been dismissed without serious consideration.

The short-range repulsion is as big a mystery as the attraction. Why does a force suddenly stop like that? One would think that a force that strong would continue to dominate the other forces until the nucleons are in virtual contact with each other. It is often said that the Pauli Exclusion Principle, named after Wolgang Pauli, is responsible, but the exclusion principle is a statement that two identical particles, like protons, cannot have the same state. This principle, however, does not attempt to describe an underlying force component that comes into play to push the identical particles apart when they are close together. Additionally, if two protons have different spin states, $-\frac{1}{2}$ and $+\frac{1}{2}$, they are not technically in the same state, and yet they are still repelled. The Pauli Exclusion Principle is an unsatisfactory answer.

Hideki Yukawa

In December 1930, Wolfgang Pauli sent a letter to some colleagues in which he relayed that he had come upon the idea that there must be an additional stable, electrically-neutral particle to conserve energy and momentum during beta decay, such as when a neutron decays into a proton and an electron, the beta particle. The problem is that the free electron has been seen to have a continuum of possible energies, so the total energy of the decay process is much greater than can be explained by the electron energy alone. The extra energy was not carried by photons, so it had to go someplace else. A few years later in 1934, Enrico Fermi published his theory of beta decay, which included the neutrino in the theory as something of a mediating particle that carried the excess energy and momentum.[128]

The idea that an interaction could be mediated by an additional particle inspired Hideki Yukawa to apply the same basic principle to the mysterious strong nuclear force. In his 1935 paper, Yukawa speculated that the strong force was mediated by a particle of medium energy.[127] It was not very long before such medium energy particles, which became known as mesons, were found, with the first being the pion. Almost immediately, Yukawa's theory was accepted and the strong nuclear force was said to be due to the exchange of neutral pions. A single pion exchange event, however, was found to not be adequate to describe the interaction in its entirety, especially not the repulsive force. Additional exchanges were theorized, such as two pion and omega particle exchanges. With the advent of quark theory, these exchanges were now understood to be meson exchanges were the mesons where composed of quarks.

A neutral pion is composed of the quark arrangement specified in Equation 25-1. Note how a down and anti-down quark pair is subtracted from the up and anti-up quark pair, and the result of the quark pair subtraction is divisible by the square root of two. It is not clear how one fundamental particle is subtracted from another fundamental particle or how a fundamental particle is divisible by the square root of two. The quark model for the neutral pion clearly fails the numerology test. Incidentally, it also fails for six other mesons where their similar equation adds or subtracts fundamental particles from each other while dividing them by the square root of two, three, or six.

Equation 25-1

$$\pi^0 = \frac{u\bar{u} - d\bar{d}}{\sqrt{2}}$$

While the particle exchange model for the strong nuclear force interaction has physicists nodding their heads in agreement like they think they know what is going on, there is no clear understanding of what is pushing on the proton to make the strong force happen and why. It also does not explain how that force becomes repulsive at short distances. The current model for the strong force does not meet the test standards for a force in the zero-point universe.

What is Really Going On?

To better understand the strong force we must go back and look at a reasonable model of a proton. Based on the shell diameter-to-mass relationship, it is likely a semi-spherical shell with an open structure. For our purposes, we can imagine it to be something like a rhombic triacontahedron. It may turn out to be something quite different, but the important thing for the purposes of this discussion is that it be a quasi-spherical shell with a given mean radius, and able to exclude zepton wavelengths equal to the shell diameter. At the same time it must be transparent to shorter zepton wavelengths. Figure 25-1 shows two hypothetical particles a distance *L* apart where *L* is slightly more than the diameter of the particles, about where the strong force begins to overcome Coulomb repulsion.

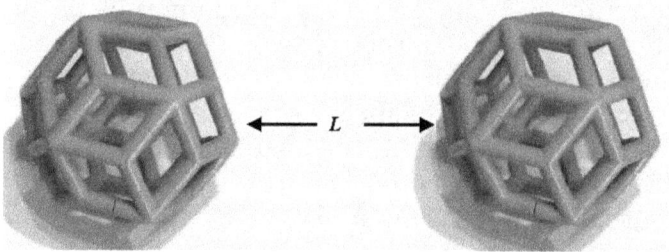

Figure 25-1 Two hypothetical particles depicted as rhombic triacontahedrons a minimum distance *L* apart where *L* is similar to the diameter of the particles. [Pictured with permission, Roger von Oech's Star-Ball® www.CreativeWhack.com]

Looking at the picture, it appears that we have a situation that is a cross between Fatio's Push Gravity plus other van der Waals Forces along with the Casimir Effect. The transparent nature of the particles to shorter wavelengths means that the shorter wavelengths will not push on the particles very effectively. Basically it looks like the Casimir-van der Waals Force should be

260

responsible. We need to keep in mind that at this small a distance, the full range of van der Waals Forces are in play, from the long-range $1/r^2$ varying force to shorter-range London-van der Waals Forces.

The maximum attraction due to the Casimir Effect occurs when a significant number of wavelengths smaller than the outer diameter of the proton begin to be excluded. Going back to Chapter 16, the outer diameter of a proton was computed to be **1.924 fm**, which corresponds closely with the maximum effective range of the Strong Nuclear Force. The inner diameter was computed to be **1.586 fm**. The Casimir Effect will continue to the point where the proton structure is transparent. Based on the rhombic triacontahedron model, the ratio of the edge length to the mean radius is **1/1.5457**, so the edge length of a proton with a mean radius of **0.8775 fm** is **0.57 fm**. The dimensions of each rhombus that makes up the triacontahedron is **1.05** times the edge length across the small dimension and **1.70** times the edge length across the long dimension, or **0.60 fm** and **0.97 fm** respectively. That means that a rhombic triacontahedron with an average diameter equal to that of a proton would become transparent to wavelengths of **~0.7 fm**, and that point Casimir-van der Walls forces would no longer be effective, allowing Coulomb forces to dominate.

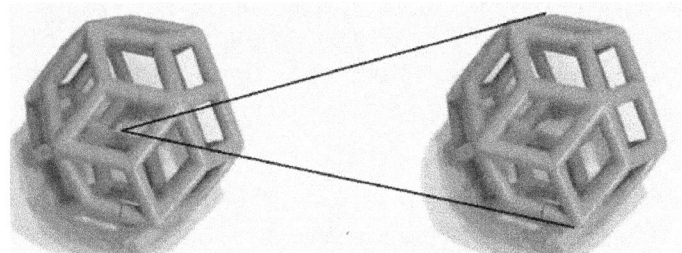

Figure 25-2 Two lines show the van der Waals Force shadow cast by one particle on another, again depicted as rhombic triacontahedrons. [Pictured with permission, Roger von Oech's Star-Ball® www.CreativeWhack.com]

Going back to Fatio's Theory we can consider a van der Waals shadow cast on one proton by another as illustrated in Figure 25-2. It is easy to see that if a force is transmitted linearly between particles that are transparent to smaller wavelength, the force will diminish when the openings in the particle structure are smaller than the shadow. In the above image, it is easy to see that the shadow in this case is slightly smaller than the smaller dimension of the rhombus (**0.60 fm**) when the particles are separated by ~ **1** particle diameter, ~**2 fm**. The van der Waals Force pushing two such particles together is strongest when the shadow is larger than the openings in the particle structure. The range of the Strong Nuclear Force at both the maximum and minimum is exactly what would be expected if protons have similar dimensions to the rhombic triacontahedron model, assuming the van der Waals Force is strong enough to account for it.

That is not to say that the structure of the proton is a rhombic triacontahedron, at least not with any degree of certainty, but rather that the domination of Coulomb repulsive forces at shorter distances can be attributed simply to the proton being transparent to shorter wavelengths. The transparent limit for the particle corresponds to the distance where the Strong Nuclear Force diminishes. Likewise the onset of the Strong Nuclear Force correlates with the same porosity of particle structure. The question remains, however, is the notoriously weak Casimir-van der Waals somehow strong enough to become the Strong Nuclear Force over short ranges?

A First Approximation
A search of the Internet reveals a paper by Ardeshir Mehta, which contains a simple computation of the Casimir Effect at 1 fm, using Equation 25-2.[129] Equation 25-2 is the standard Casimir Force equation for two flat plates of area **A** and distance **L** between them to which Mehta added a modifier **x**. He arbitrarily assigned a value of **x = 5** in order to correct for the curvature of the nucleons. It has been rewritten here slightly for

consistency. The negative sign indicates that the force is attractive.

Equation 25-2

$$F = -\frac{\hbar c \pi^2}{240x}\frac{A}{L^4}$$

He obtained a value for the Casimir Force of **F = 1.04 x 10³ Newtons (N)**. This compared to his computation of Coulomb repulsion of **F = 31.64 N**. Based on his calculations, the Casimir Force was almost **33 times** the Coulomb repulsion at **1 fm**, which is in fairly good agreement with the Nuclear Force strength. Note that there was a simple error as he used twice the approximate area of the spheres in the calculation when that factor of **2** is already included in the equation (240 instead of 480 in the denominator), so his result for the Casimir Force should have been **16.4 times** the Coulomb repulsion. The result is nonetheless encouraging.

A Better Approximation

A better Casimir Effect energy equation for two spheres can be found in Appendix A of a paper by Aurel Bulgac et. al.[130] Their energy equation for two spheres with a common radius **a** and minimum distance between them **L** is shown in Equation 25-3. Note that this equation was derived by what is known as the proximity force approximation or PFA. This approximation treats each little part of a non-flat object as being parallel to the opposite body. Basically each piece is under the same force as modeled with the standard two-plate equation. This is one way to compute the Casimir Effect between two curved surfaces, such as two spheres, that otherwise might not be possible to compute. It is unknown how good this approximation is in this instance, but it gives us a point of comparison.

Equation 25-3

$$E = -\frac{\hbar c \pi^2}{1440} \frac{\pi a^2}{L^2(2a + L)}$$

We can note that **a** is the radius of the proton, which is effectively constant. It is also similar in magnitude to **L**, such that we could treat it like a multiple of **L** and solve for the energy. If, for example, we start with **2a = L = 1.755 fm** such that the minimum distance between the protons equal to the proton diameter, we get Equation 25-4.

Equation 25-4

$$E = -\frac{\hbar c \pi^2}{1440} \frac{\pi}{8L}$$

To convert from an energy term to a force term, Equation 25-4 must be differentiate and multiply by **2**. This gives a resulting force Equation 25-5.

Equation 25-5

$$F = -\frac{\hbar c \pi^3}{5760} \frac{1}{L^2}$$

That can be compared to the Coulomb repulsion computed by Equation 25-6.

Equation 25-6

$$F = \frac{1}{4\pi\varepsilon_0} \frac{q^2}{r^2}$$

Note that in this special case **r = 2L**. The Casimir Force pushing the two protons together is **55.3 N**, while the Coulomb Force pushing the two protons apart is **20.6 N**.

The Casimir Force is **2.69** times stronger using Equation 25-3 at a minimum approach distance of **1.755 fm**. This distance is greater than calculated by Mehta, and the ratio is lower, but the net of these two basic forces is attractive at this distance.

Computation
In order to get a better idea of the force magnitude over the entire effective range using the above the Casimir proximity force approximation equation, calculations were done for minimum distances L in **0.1 fm** increments from **0.1 fm to 4 fm** can be done. The Coulomb repulsion was also computed using the proton charge and the distance between the center of the spheres based on the same distance L. Lastly the ratio of the of the Casimir Force over the Coulomb Force is determined. The results are shown in Chart 25-1.

In order to correct for the disappearance of the Casimir-van der Waals Forces at the shorter distances, the force between **0.1 fm** and **0.5 fm** was manually set to zero. The force was also manually adjusted to approximate a linear decline in strength between **0.9 fm** and **0.5 fm**. This was done for purposes of illustrating the effect due to the semi-transparent structure of the proton. The ratio value was multiplied by **10** so that it would show up on the chart. The peak ratio is **15.1** at **0.9 fm.** This is in close agreement with Mehta's approximation.

Discussion
It is clear from these approximations that the Casimir Effect is no longer weak at femtometer ranges, but becomes a Strong Casimir Force. If there were not some mechanism whereby the Casimir Force goes away at short ranges, the Strong Casimir Force would reach **>50,000 N** using this equation. To put that into perspective, that force may be greater than the gravitational attraction at that distance from a proton sized black hole.

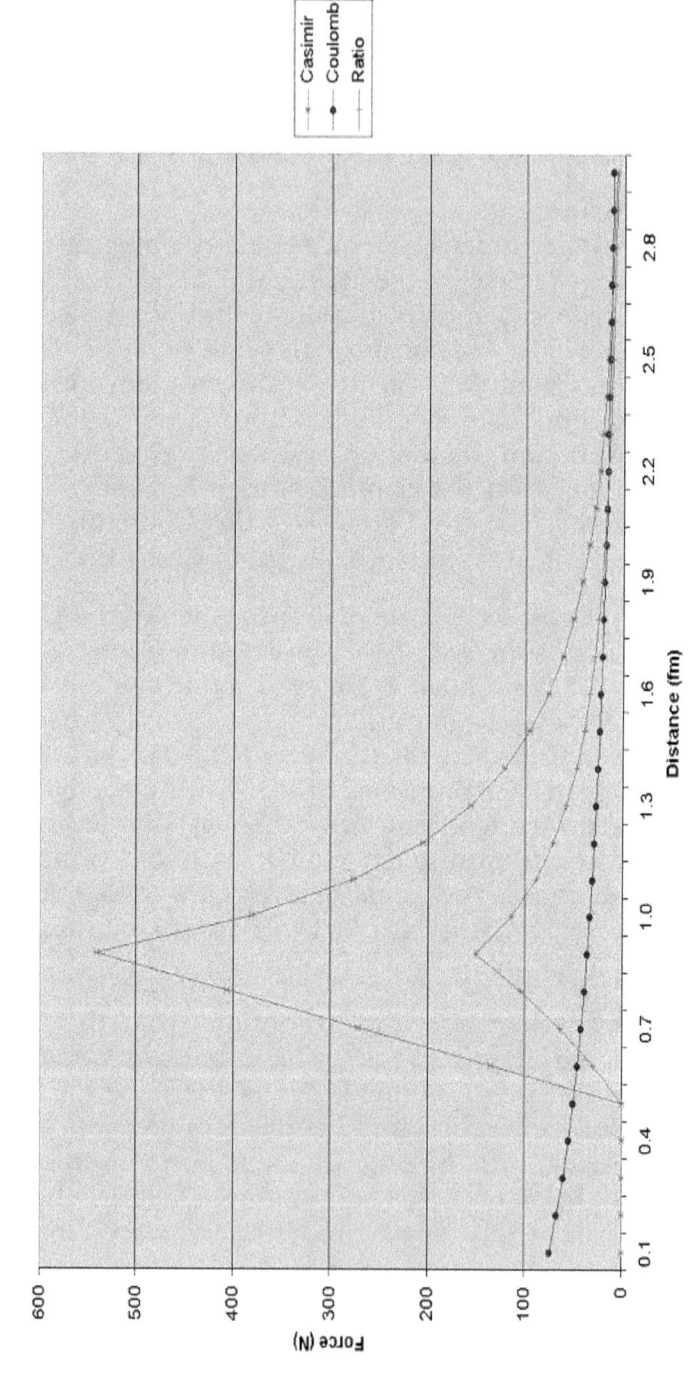

The Casimir Force becomes stronger than the Coulomb repulsion in the above chart at a distance slightly more than **2.6 fm** and becomes twice the Coulomb repulsion at a little less than **2.0 fm**. These distances correlate well with the Strong Nuclear Force for an approximation. If we were to include the effect due to the van der Waals Force shadow size relative to the openings in the proton structure, the distance where the Casimir Force overcomes Coulomb repulsion would be closer to **2 fm**.

The main outstanding issue with the Strong Casimir Force model is the magnitude, as the Strong Nuclear Force is generally accepted to be 100 times greater than Coulomb repulsion and in this approximation it is only 15 times greater. This equation was not derived to account for Casimir Forces on a semi-porous quasi-spherical shell, and may not account for all the van der Waals pressure forces pushing the nucleons together. The equation also only takes the particle area into account rather than the entire mass-energy. The proximity force approximation may also not be very accurate when the distance between the spheres is similar to their diameter. There is enough unknown about the computation that it is entirely possible that the Strong Casimir Force can account for the entire Strong Nuclear Force between nucleons.

Could there be two forces that act over exactly the same range with a similar magnitude? While certainly possible, it seems unlikely. The Strong Casimir Force has several advantages over the existing alternatives. First of all it is based on London-van der Waals Forces, which are well known and act in a simple manner. The Casimir Effect has been proven experimentally at the micron range, so its existence cannot be dismissed. There is no known mechanism by which the Casimir Effect could be thought not to act at the femtometer distance range, so it is at least responsible for part of the Strong Nuclear Force intensity. Probably the most important point in favor of the Strong Casimir Force model is that the Casimir Effect is part of Quantum Electrodynamic theory, so the Strong Nuclear Force no longer needs to be thought of as a separate force, and can be unified with the Electro-Matter Force.

The Cut-Off Wavelength
Early on the idea of a cut-off wavelength was introduced as a means of capping the total energy of the vacuum, to prevent it from being infinite. The cut-off wavelength was thought to be the Planck Length. Then we found that the vacuum pushes on both sides of bodies of matter more or less equally, making the cut-off limit unimportant. The idea that particles are completely transparent to shorter vacuum fluctuation wavelengths now makes that the concept completely unnecessary, as the effective cut-off in vacuum pressure occurs when particles, in particular protons and neutrons become transparent. Higher energy shorter wavelength zeptons simply go through particles without interacting.

Conclusion
While the Casimir Effect is usually thought of as a very weak force, computations show that it becomes very strong at distances of a few femtometers or less. This Strong Casimir Force is entirely consistent with the known distance range of the Nuclear Force within the atomic nucleus. While the magnitude computed herein is less than the known magnitude of the Strong Nuclear Force, there are a number of adjustments that need to be made in order to compute the force more exactly. These adjustments will in all likelihood yield a force of the correct magnitude.

It seems unlikely that both the current particle exchange model and the Strong Casimir Force could both be true, so the ultimate question is then should the particle exchange model be discarded? The particle exchange model certainly has time in its favor, as most physicists were indoctrinated into the theory. The particle exchange model does have its weaknesses, particularly that it is not in accordance with zero-point field theory. The Casimir Effect, as a London-van der Waals Force, has a strong fundamental basis. The Casimir Effect is also the simpler and more elegant of the two solutions, and ultimately allows us to unify the Strong Nuclear Force with electromagnetic theory, thus eliminating an unnecessary "fundamental" force. We can only conclude that the Nuclear Force is the Strong Casimir Force and part of the Electro-Matter Force.

The new points we can add to the zero-point universe are:

77) The Nuclear Force is the Strong Casimir Force
78) The Nuclear Force is a part of the Electro-Matter Force

[127] H. Yukawa, "On the Interaction of Elementary Particles," PTP, 17, 48, 1935.

[128] E. Fermi, Z. Physik 88 161, 1934.

[129] A. Mehta, "Modification of the strong force nuclear force by the zero-point field," April 7 1999.

http://homepage.mac.com/ardeshir/ModificationOfTheStrong....pdf

[130] A. Bulgac, P. Magierski, A. Wirzba, " Scalar Casimir effect between Dirichlet spheres or a plate and a sphere," Phys. Rev. D 73, 025007, 2006. doi:10.1103/PhysRevD.73.025007

Chapter 26: The Weak Interaction

The more you see how strangely Nature behaves, the harder it is to make a model that explains how even the simplest phenomena actually work. So theoretical physics has given up on that.[131]

Richard Feynman, 1985

Contrary to Feynman's lamentation, sometimes the simplest phenomena have the simplest of explanations. Such is the case with the weak interaction. Note in particular that even though it is called a force within the Standard Model, it is more accurately just an interaction. It is not even called a force within the current quark based model. To start with we should review the current model, because it has a number of difficulties.

W and Z Particle Theory
The W$^+$ and W$^-$ are known to have very short half-lives, ~10^{-24} seconds per most common literature. However, if it they were virtual, so that energy could be conserved during an interaction event, it would be limited to **1.29 x 10^{-26}** seconds per Heisenberg's Uncertainty Principle. If we look at the maximum wavelength, which equates to a virtual **80,398 MeV** W$^+$ or W$^-$, we get a length of **3.855 x 10^{-18} meters**. The radius of the proton is **0.8775 x 10^{-15} meters**, so the W particles are ~**230** times smaller than the radius of a proton, and similarly smaller than a neutron. It would appear that in the vast majority of cases the W$^-$ would decay well inside the proton, and the electron would be re-captured rather than emitted. Even if an electron is released at the surface of a proton, it seems improbable that the electron could get far enough away to avoid being recaptured unless the W particle gives it a large amount of energy. However, if the W particle is a virtual particle and it gives up energy; that violates the principle of conservation of energy. Actually if a virtual particle decays into an electron then it has already violated that principle.

The entire W particle theory suffers any way you look at it, since if it is produced from a quark and exists for more than **1.29 x 10⁻²⁶** seconds, it violates conservation of energy and the Uncertainty Principle. If it cannot exist longer than that, it fails to describe the phenomena. In order to not violate conservation of energy, the W particle must be virtual. In order for it to be virtual it must be produced as a pair and it must annihilate within the time limit stated above. If is a W-antiW pair, which is not in accordance with the theory, it could hypothetically be produced from the vacuum and then annihilated. But if that is the case it could not decay into a lepton and a neutrino, as the theory requires, since all its energy must be returned to the vacuum. If a virtual W or Z is composed of another unidentified particle pair, which is also not part of the theory, then it still cannot decay into something else without violating conservation of energy. Any other production of a meta-stable **80,398 MeV** particle violates the principle of conservation of energy, since it is much more massive than the particle which is supposed to be producing it.

There is also the difficulty of why beta particles, radiated electrons, have a continuum of energies. In what we might perceive as a normal particle interaction, a certain fixed amount of energy would be released during beta decay, so ideally we would expect for the beta particle to always have a fixed amount of energy, not a continuum of possible energies. The W and Z model does not adequately address how the energy range comes about. The neutrino assumes the role of the particle needed to carry away the energy difference, but that does not address why there is an energy continuum in the first place.

Even if the W and Z model for beta decay is accepted, there must be a second mechanism to allow an electron to gain an appropriate separation distance from the proton or otherwise, a source of energy, which does not violate energy conservation principles. It seems more likely that the W and Z particles as gauge bosons theory is just as nonsensical as the photon being a gauge boson for electromagnetic theory, if not more so.

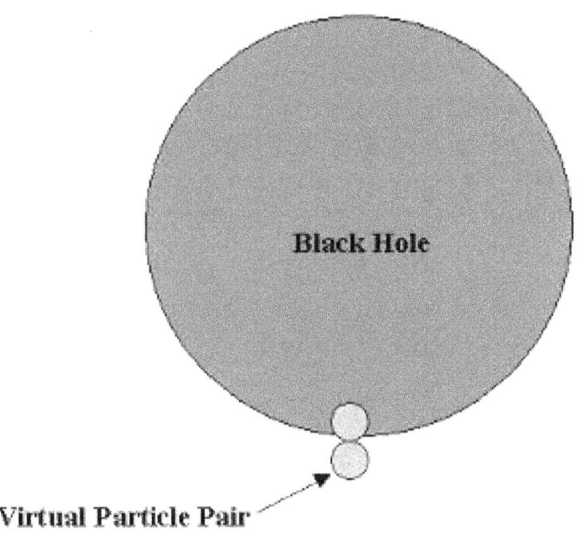

Figure 26-1 An illustration of Hawking Radiation where a virtual particle pair is formed just outside the event horizon of a black hole and half of it falls inside.

Hawking Radiation

OK, you are probably thinking that Hawking Radiation is coming from left field, but bear with me. Hawking Radiation occurs when a virtual particle pair, a vacuum fluctuation, occurs just outside the event horizon of a black hole and one half of the pair enters the black hole, while the other remains outside.[132] This effect was first proposed by Stephen Hawking, and while it has yet to be experimentally verified, it is a commonly accepted theory. The attractive force of the black hole is too strong for the particle pair to recombine, so the particle outside the event horizon becomes free, while the one inside is confined within the black hole. The black hole has to give up energy equivalent to the particles' pair production energy, some kinetic energy, and the energy of some photons in order to conserve energy and momentum.

273

Neutron Decay

Neutron decay is perhaps the simplest weak interaction, a form of beta decay, which occurs most frequently in free space. While neutrons inside stable atoms appear to exist indefinitely, free neutrons are known to decay with a half-life of slightly more than 10 minutes. The term "half-life" means that half of them decay in that amount of time. The rapidity of this reaction is consistent with an interaction that occurs fairly regularly.

Neutrons are formed when an electron is accelerated toward a proton or *vice versa*, with enough energy to overcome the repulsive force that is normally present in a hydrogen atom, which prevents an electron from falling into a proton. If the neutron is in free space, it quickly decays, releasing the electron and leaving behind a proton. Since electrons are also known as beta particles, this is considered a form of beta decay. For the purposes of this discussion, it is useful to think of a neutron as being a proton collocated with an electron.

As with any particle, a neutron is surrounded by and filled with zeptons. What if one of those zeptons is an electron-positron type and the positron half of it crosses into the neutron's shell while the electron remains outside? And say that it has enough energy for electron-positron pair production to occur, although that may not be a requirement. What we then have are perfect conditions for something similar to Hawking Radiation to occur. The virtual positron combines with the electron inside the neutron, annihilating with it, and leaving behind a proton, while the virtual electron is freed at some distance away from the proton. This is the manner in which a neutron decays in free space.

Electron Capture

A very common form of radioactive decay of atomic nuclei is electron capture. In the case of electron capture, one of the atom's orbital electrons is captured by the atom's nucleus. The electron combines with one of the protons of the nucleus forming a neutron. This does not change the total mass of the nucleus very much, but by having one fewer proton, it is converted to the next lower atomic number element. In the process, some photons are emitted including one from another orbital electron, which takes the place of the one that was captured.

There has always been a question about how an orbital electron is captured, since it would not normally come so close to the nucleus so that it could be snatched up by a proton. There needed to be a transport mechanism allowing the electron to jump from orbit into the nucleus, overcoming the repulsive force. The minimum zepton energy required to produce an electron-positron pair, **1.022 MeV**, equates to a wavelength of **6.07 x 10^{-13} meters**. As long as an electron-positron pair is being annihilated, it may not even need to have that much energy, so a less energetic virtual electron-positron pair with say a **10^{-11}** meter wavelength could easily facilitate electron capture. In other words, the less energetic zepton could easily snatch an electron from its normal probability zone by annihilating it with a positron and depositing the once virtual but now free electron in the nucleus where it is absorbed by a proton.

Positron Emission
Positron emission is a lot like electron capture except that the virtual positron does not annihilate with an orbital electron. Instead, the positron is given so much energy that it is not immediately annihilated by an orbital electron and so escapes the atom, radiating into space. It eventually loses its energy interacting with other matter and eventually does annihilate with an electron. As with other forms of beta decay, this interaction is mediated by an electron-like zepton.

Beta Decay
While the last two decay methods involved nuclei increasing the number of neutrons, beta decay acts in the opposite manner, decreasing the number of neutrons for when a nucleus already has too many to be stable. If we consider the above Hawking Radiation mechanism, we can conclude that a virtual electron-

positron pair will at some point be produced in the vacuum next to the nucleus and the positron side will enter a neutron. The electron inside the neutron then annihilates with the positron, leaving a free electron at some distance from the neutron with enough excess energy that it becomes free of the atom and radiates into space.

Virtual Proton-Antiproton Interactions

All of these reactions could be thought of in terms of virtual proton-antiproton interactions. For example, if an antiproton from a proton-antiproton pair interacts with a neutron, the part of the neutron that equates to a proton will be annihilated, leaving behind a free electron and a proton deposited off to the side. To get an idea of the range, the energy of proton-antiproton pair at the pair production energy is **1.877 GeV**, which equates to a wavelength of **0.33 x 10^{-15} meters**. This is less than the radius of a proton but much longer than the W and Z particle wavelengths. We can have one of two things happen when proton-antiproton initiated beta decay happens. Either the freed electron recombines with a proton inside the nucleus, likely the same one, forming a neutron again and maintaining the current form of the nucleus, or the free electron is ejected from the nucleus. The latter case is one mechanism that could explain a second beta decay energy and account for those atoms with two modes of beta decay. If the electron is recaptured then it will make the nucleus look like it is vibrating, while a lot more is really happening.

The other forms of beta decay can also be explained by a proton-antiproton Hawking Radiation mechanism, assuming the freed particle is not recaptured in all circumstances. And, if there is such a thing as neutron-antineutron pairs, then they too could take part in weak force interactions.

Beta Decay: Continuum of Energies

Interestingly, the Hawking Radiation Weak Interaction also gives us an explanation for why beta particles and positrons have a continuum of energies. The zeptons that are responsible for the Weak Interaction have a continuum of energies, which leads directly to the beta and positron emissions having a continuum of energies. The original zepton must have energy within a specific range in order for the interaction to occur, and there must be a probability function with respect to zepton energy such that zeptons with certain energies are more likely to cause beta decay than at other energies. A probability function is necessary to explain why the energy distribution of the beta and positron particles is not uniformly flat. Interaction probabilities are also necessary to explain the wide variety of half-lives of radioisotopes.

Neutrinos

Neutrinos were largely left out of the previous sections since they are not an immediate part of the weak interaction mechanism, but rather are required in order to conserve energy and momentum at the end of an interaction. The neutrinos or antineutrinos are emitted after an interaction per the usual formulæ.

The precise form of the neutrino within the scope of the zero-point universe theory is not clear at this time. There is, however, one idea, which at the moment, I consider to be highly speculative, but worthy of inclusion. If we consider the form of the photon discussed in the beginning, it is composed of a series of counter-rotating central zeptons. There should be two other forms of photon: one where the zeptons always rotate forward and one where the zeptons always rotate backwards, without counter-rotating. The forward or backward direction is determined by whether the matter side of the virtual particle pair, such as the electron, is moving forward or backward time. Such photons would have an **h/2** spin and meet many, possibly all the properties of neutrinos.

Other than that neutrinos appear to simply be a fundamental mechanism for the zero-point field to transmit energy, in a slightly different manner from a photon. A more in depth discussion will be saved for another day and perhaps another book.

Conclusion

Hawking Radiation leads us to a logical and intuitive mechanism that explains beta decay, electron capture, positron emission, and neutron decay. Hawking Radiation allows a virtual particle pair to cross into a nucleon and annihilate part of a neutron or produce a neutron. This is a simpler and a more fundamental way of explaining the weak interaction than the W and Z model in existing theory. As it turns out, the weak interaction is not a force at all and should not be considered one of the fundamental forces. Even so it is entirely consistent with Electro-Matter Force Theory and the zero-point universe. Since it has been established that gravity and the strong force are part of the Electro-Matter Force and the weak force is not a force at all, that leaves us with only one force: the Electro-Matter Force.

The following points are now clear:

79) The Weak Interaction is due to Hawking Radiation
80) The Weak Interaction has an electro-matter origin
81) There is only one force, the Electro-Matter Force

A Few Afterthoughts

On a somewhat more speculative note, one might also consider the Hawking Radiation beta decay mechanism with respect to orbital electrons. What happens when an electron-positron-like zepton annihilates with an orbital electron and does not deposit it in the nucleus? Of course, the newly freed, once virtual electron will be deposited some place near the nucleus, still in orbit. What if this happens repeatedly, many times per second? Then we would have a probabilistic cloud of electron locations, matching our observations.

What then if a lower orbital is vacant? The electron can be instantaneously deposited into a lower energy orbital, with a photon emitted after the fact to equalize the energy. It has always been a puzzle about how an electron can make instantaneous jumps between orbitals, but it becomes simple, even obvious, once the Hawking interaction mechanism is understood. So not only does the Hawking Radiation model provide a simple explanation for beta decay, it also provides a simple semi-classical model that bridges the gap between a classical Bohr type atomic orbital model to a quantum mechanical orbital model.

Similarly, as previously hinted at, proton-antiproton pairs within the nucleus would annihilate with protons, or the proton-like part of a neutron, caused the nucleus to appear to resonate very rapidly. To take it even further we can consider that all motion at the quantum level is mediated by zeptons with every movement requiring billions of annihilation and production events per second for every particle in our body.

[131] R. P. Feynman, QED: The Strange Theory of Light and Matter, Princeton University Press, Princeton, New Jersey 1985

[132] S. Hawking, "Black hole explosions?" Nature 248, 30 (1974). doi:10.1038/248030a0

Chapter 27: Conclusion

Must have something to do with the zero-point energy.[133]

Neils Bohr, ~1946

A Few Thoughts

Hendrik Casimir related a conversation he once had with Neils Bohr in 1946 or 1947, where after Casimir described his work on what became known as the Casimir Effect and his efforts for finding a simpler way of deriving it, Neils Bohr suggested that it "must have something to do with zero-point energy."[133] If we had to pick a moment in time when zero-point field theory began anew, perhaps that was it. It was the Casimir Effect that shone the light on a roadmap for physics that includes an underlying mechanism for force transmission incorporating zero-point energy into a force, rather than renormalizing it out of the theory.

This came after a dark period in time when zero-point energy was largely treated as a curiosity, or a difficulty. It is unfortunate that so many physicists had an excellent understanding of how mechanical force transmission must work with respect to an æther that fills the vacuum of space in the mid to late 1800s, Maxwell's vortex model of magnetism and Lord Kelvin's dynamical spinning motion to explain inertia to name a few. All that understanding along with the mechanism behind the transmission of light was swept aside when those who denied the æther theory took control. Sadly, it was only decades away from when Dirac was able to fundamentally describe the zero-point field with a polar particle pair model, as by then, nobody was willing to listen. Worst than that, physicists chose to close the book on theories and close their minds to new possibilities while leaving the question of the underlying force mechanics unanswered. It has been a sad century for fundamental physics, a true dark age. At the same time, it has been a great century for physicist-engineers who have made great advancements. There is no telling what might have been accomplished if there was a sound fundamental foundation.

Theoretical physicists should never have given up their efforts to describe the fundamental nature of the universe, but rather should have adopted a more stringent attitude with respect to requiring that physical models of the universe comply with certain immutable laws. The first law is that every force is transmitted point-to-point through space. That point-to-point transmission requires that the medium of space be involved in every interaction. That medium is the zero-point field. To put it more succinctly, every force is transmitted by and through the zero-point field. Once that criterion is in place, we find the possibilities limited to such a degree that there is only one possible set of conclusions, the ones outlined here in this book.

Physicists must not be sloppy in their implementation of force fundamentals as they have been for more than a century. Once the fundamental sloppiness is eliminated, the possible answers to problems are severely limited. In almost every case we have examined, the choices have been reduced to a single possibility, and those possibilities add up to a self-consistent whole. That is not to say that I have not been sloppy with the details, but if one must be sloppy, it is better to do so with details than with fundamentals. If the fundamentals are correct, the details can be worked out. Or, in the spirit of Dirac, get the class-one problems right, and then the class-two problems can be readily worked out by the legions of scientists. If the class-one problems are answered incorrectly, the class-two problems can never be solved.

The Zero Point Universe
So what is the zero-point universe? It is a universe filled with vacuum fluctuations, the zeptons. The zeptons carry positive and negative electric charge and matter and antimatter matter charge. All force interactions can be attributed to electro-matter charge interactions following the principles of the Electro-Matter Force. All

basic forces can be described under the fundamental force theories of Maxwell, van der Waals, and Casimir, and those who contributed to their development. There is only one force, the Electro-Matter Force.

Furthermore, even the stable matter, the electrons, protons, and neutrons, appear to be virtual in nature, being composed of either zeptons or the material that makes up the zeptons. Mass as we have seen is equivalent to the zepton energy excluded by those particles. Everything in the zero-point universe is virtual.

That, my friends, is the zero-point universe, at least the fundamental forces. As for the rest, in particular, particle theory, cosmology, or the composition of zero-point energy, I can only say that many of my ideas on those are speculative and at best not ready for publication. Perhaps in the future they will be topics for additional books. So now you are thinking: the ideas in this book are not speculative? My response is that after more than a decade living with the reality of Electro-Matter Force Theory, I have become comfortable that all the ideas presented are factually correct, except where more speculative ideas are noted. Certainly there are many oversimplifications, but when all is said and done, the best a physicist can hope for is that his or her theory will be an oversimplification, as other physicists probe for, and eventually acquire, a deeper understanding of the physical universe.

For the rest of you, those who continue to deny the existence of zero-point vacuum fluctuations, all I can say is, good luck with that.

[133] H.B.G. Casimir, Some remarks on the history of the so called Casimir effect, The Casimir Effect 50 Years Later, Rd. M. Borlag, World Scientific Publishing Co. Pte. Ltd. (1999) pg 6

www.ingramcontent.com/pod-product-compliance
Lightning Source LLC
Chambersburg PA
CBHW051444170526
45166CB00001B/109